Julie -
May all your e-ventures
be a resounding success!

Shari

e-Business in Manufacturing

Putting the Internet to Work in the Industrial Enterprise

e-Business in Manufacturing

Putting the Internet to Work in the Industrial Enterprise

Shari L.S. Worthington
Walt Boyes

ISA–The Instrumentation, Systems, and Automation Society

Notice

The information presented in this publication is for the general education of the reader. Because neither the author nor the publisher have any control over the use of the information by the reader, both the author and the publisher disclaim any and all liability of any kind arising out of such use. The reader is expected to exercise sound professional judgment in using any of the information presented in a particular application.

Additionally, neither the author nor the publisher have investigated or considered the affect of any patents on the ability of the reader to use any of the information in a particular application. The reader is responsible for reviewing any possible patents that may affect any particular use of the information presented.

Any references to commercial products in the work are cited as examples only. Neither the author nor the publisher endorse any referenced commercial product. Any trademarks or tradenames referenced belong to the respective owner of the mark or name. Neither the author nor the publisher make any representation regarding the availability of any referenced commercial product at any time. The manufacturer's instructions on use of any commercial product must be followed at all times, even if in conflict with the information in this publication.

Printed in the United States of America.
10 9 8 7 6 5 4 3 2

ISBN 1-55617-758-5

ISA—The Instrumentation, Systems, and Automation Society
67 Alexander Drive
P.O. Box 12277
Research Triangle Park, NC 27709
www.isa.org

Library of Congress Cataloging-in-Publication Data

Worthington, Shari L. S.
E-business in manufacturing :putting the Internet to work in the industrial enterprise / Shari L.S. Worthington, Walt Boyes.
p. cm.
Includes bibliographical references and index.
ISBN 1-55617-758-5
1. Manufacturing industries–Technological innovations. 2. Electronic commerce. I. Boyes, Walt. II. Title.
HD9720.5 .W64 2002
658'.054678–dc21

2001005261

Dedication

To my husband, Mark, and my daughter, Nikki, for their love, encouragement, and patience. To my parents, Carolyn and Floyd, who raised me to work hard, be honest, and persevere come what may. To my sister, Linda, who's always been there for me. – Shari

To my wife, Betsy, and my daughter, Andrea, who put up with my "disappearing into the computer" for months on end and still encouraged me to keep on with it. – Walt

Acknowledgements

Now that this tome has gone to press, we're pleased to report that the book writing process isn't quite as difficult as we first imagined. In our case, that's because we had the help of a number of generous and capable people. We are deeply grateful to our publisher, Bob Burns, and his able team of copy editors, graphic designers, and publicists. We are also thrilled to have been the recipients of an incredible amount of insightful feedback from industry analyst extraordinaire Dan Miklovic, pricing guru Jim Geisman, control community maverick Ken Crater, automation industry pundit Dick Morley, futurist Jim Pinto, and Industrial Marketing Practice Association President Prentis Hall. Our heartfelt thanks also to those who agreed to allow us to interview them for this book: Deborah Vollmer Dahlke, Fred Putnam, Karen Turner, Don McCook, Julie Harrington, Jacques Werth, Darron Smith, Chuck Steinberger, Charlie Gifford, Jeff Cawley, Bob Hitomi, Dennis Brandl, and anyone else we mistakenly forgot to mention. Their willingness to share their experiences with the rest of the manufacturing industry is a tribute to the strength of this community.

We'd also like to thank the powerful woman behind the scenes of the ISA Publications Department, VP-Elect Diana Bouchard. She dragged us into this project because she knew it was the right thing to do. We kicked and screamed a lot, but we finally acquiesced. Turns out she was right and we're pleased to count her among our friends.

Shari would also like to express her deep thanks to her staff, especially Linda, Frank, Jerry, and Tom, who picked up extra shares of the Telesian workload while this book was coming together.

About the Authors

Shari L.S. Worthington

Shari Worthington is President of Telesian Technology Inc. After a decade in increasingly senior product management roles, Worthington established Telesian in 1987. The firm provides industrial marketing and Web development services to the manufacturing industries, working with such clients as Raytheon Co., Apple Computer, and Dresser Instruments. A scientist turned engineer turned marketeer, Worthington's specialty is real-time computing in laboratory and manufacturing applications. She holds a BS in biology and psychology from St. Lawrence University, an MA in psychology from Framingham State College, and an MBA in marketing from Babson College.

An entrepreneur at heart, Worthington has been active on the Internet since the mid 1980s. In past lives she founded MacSciTech, an international association of scientists and engineers who use Apple Macinotsh computers, and launched "Science & Engineering Network News" (SENN), one of the first publications to track and analyze developments on the Internet. She later sold SENN to Nature Publishing. As a result of these ventures, Worthington has spent thousands of hours analyzing the latest and greatest technologies offered by the Net. This, combined with over 20 years of experience in industrial and technology marketing, has established her as an expert in innovative business practices, both online and off. She is an international speaker on the topics of e-business, guerilla marketing, and e-marketing.

Worthington serves on the Board of Directors of the WPI Venture Forum (www.wpiventureforum.org) and is a Past Chair. She is also a founding board member and Past President of the Industrial Computing Society (www.ics.org), a sister organization of the ISA.

Worthington can be reached at sharilee@telesian.com.

Walt Boyes

Walt Boyes is principal of MarketingPractice Consultants in Maple Valley, Washington. Walt has over twenty-five years of experience in sales, sales management, marketing, and product development in the automation, controls, and instrumentation industry, both for process instruments and environmental controls. One of Walt's proudest accomplishments is the R&D100 Award he shares with four other product development team members for development of a novel flow meter for use in wastewater. In addition, he is an authority on business-to-business marketing and on electronic commerce in the business-to-business arena. His clients include manufacturers, distributors, and electronic commerce concerns. He is also active as a technical writer and editor.

Walt is Senior Contributing Editor for *Industrial Marketing Practitioner* magazine, and Contributing Editor for *Flow Control* magazine. Walt is also co-founder, with Prentis Hall, of the Industrial Marketing Practitioner's web-seminar series. The first of the series, entitled, "Marketing on a Shoestring," debuted in September 2000. He is the Editor of *The Instrumentation Reference Book, 3rd Edition,* 2001, published by Butterworth-Heinemann Publishing.

Walt served from 1994 to 2000 as Director of Electronic Publications for ISA—The Instrumentation, Systems, and Automation Society, and as Past President of the Seattle Section, ISA. As EPubs Director, he was responsible for oversight of ISA ONLine, the online presence of the Society, as well as all other electronic publishing endeavors of the Society. He currently (1/1/2001) serves as Vice-President of Publications for ISA. He is also a member of the Industrial Computing Society, AWWA, and has served on the Technical Practices Committee for WEF. Boyes can be reached at wboyes@ix.netcom.com.

Table of Contents

Foreword

Now that the early hype about the Internet has subsided, companies all over the world are wrestling with the challenges of how to really use the Internet to transform their businesses. Manufacturing companies need real and implementable guidance on leveraging the Internet's global reach, organizational transparency, and collaborative nature. Unlike financial institutions, which deal with intangible assets, or the software industry, which can use the Internet as a primary delivery mechanism, manufacturers of real goods must integrate the electronic world of the Internet with the very real world of raw materials, manufacturing processes, and fulfillment. An entrenched axiom in manufacturing is "if it isn't broken—don't fix it." Amidst the recent wreckage of dot.coms and B2B marketplaces, manufacturers are reluctant to invest in what they see as a failed Internet business model.

The reality is that though many dot.coms and B2B marketplaces did have bad business models and deserved to fail the Internet was not to blame. Many manufacturers have used the Internet to actually transform their businesses. One of the fastest-growing segments of the electronics industry is electronic contract manufacturing (ECM), and the three largest ECM companies—each experiencing more than 30 percent annual growth—rely heavily on the Internet to enable customer relations, e-manufacturing, and supply chain management. Manufacturers, large or small, should all learn what the ECM industry has discovered: the Internet can not only transform your business; it can also transform the entire market or industry creating all new businesses.

What manufacturers need is practical user advice, with real case histories, to help them build the business case that e-business is real business. The authors of *e-Business in Manufacturing* have a narrative style that makes for an interesting as well as instructive read. They provide a refreshing blend of historical perspective with real-world suggestions and examples of how to use the Internet to drive change. Without preaching or hyperbole the authors serve as advocates for a new way of doing business in an efficient and electronically enabled way. Any manufacturer looking for insight and a historical perspective into the changes the Internet is wreaking in business will find this book an enjoyable read with enough detail to start them down the transformation path.

Dan Miklovic
Vice President
and Research Director

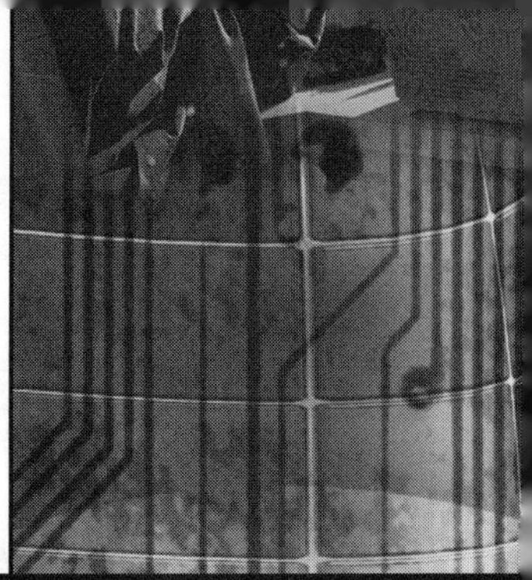

CHAPTER 1

An Introduction to e-Business in the Manufacturing Industries: Where We Are, Where We're Going

Sometimes it's hard to separate the substance from the hype, especially when it comes to the Internet. Now that the manufacturing community is finally becoming more comfortable with the Web revolution, it's time to sort through the next Internet phenomena—e-commerce, intranets, VPNs (virtual private networks), SET (secure electronic transaction), and so on—as well as the tremendous business opportunities created by the Internet.

Before we get lost in all the acronyms, let's start with the basics—what exactly is electronic business, or "e-business"? It all boils down to doing business online, finding and interacting with customers, communicating with business partners and suppliers, and developing new products and markets.

What about *e-commerce*? Does this term imply something other than e-business? The bottom line is that the term *e-commerce* focuses people's attention on the act of placing an order online when, in fact, that is only a small piece of the puzzle. We prefer to use the word *e-business* because it helps people focus on the impact of the Internet on all their business practices. And whether you're a large manufacturer of instrumentation systems or a small, in-house research lab looking for next year's funding, e-business will be an integral part of your future, in many cases sooner than you think (see Table 1-1).

Table 1-1. As shown in this example income statement, the Internet can have a dramatic impact throughout the manufacturing organization.

INCOME STATEMENT (in thousands)					**INTERNET IMPACT ON THE BOTTOM LINE**
Net Sales				$380,277	• The Internet provides access to a larger base of customers and makes ordering quick and easy. The result is increased sales and a more productive salesforce.
(note effect on A/R and A/P)					
Cost of Goods Sold					
Inventory, January 1			41,849		• Secure extranet connections allow you to put suppliers to work fine-tuning inventory to just-in-time levels, thus reducing inventory carry costs.
Purchases		173,099			
Add transportation-in		27,443			
Total cost of purchases		200,542			• Automated order placement, e-marketplaces, and EDI can have a major impact on reducing the cost of, and time required to place, orders
Less: Purchase discount	10,386				
Purchase returns and allowances	1,011	(11,397)			
Net Purchases			189,145		
Cost of goods available for sale			230,994		
Less inventory, December 31			(78,301)		
Cost of Goods Sold				152,693	
Gross Profit				227,584	
Operating Expenses					
Research & Development		38,913			• Electronic collaboration tools allows interaction between widely dispersed facilities and global partners thus increasing the productivity of R&D dollars
Sales & Marketing		70,844			• CRM and e-marketing tools allow manufacturers to develop even closer relationships with customers prospects, and partners. This results in more satisfied customers and increased sales.
General and Administrative		25,805			• Net-based information sharing tools make possible a flatter, more heterarchical organization, reducing G&A costs by removing layers of management and many clerical functions
Total Operating Expense				(135,562)	
Operating Income				92,022	
Other Income & Expense				(13,458)	
Earnings Before Interest & Taxes				78,564	

Today, integrating e-business into the manufacturing enterprise is all about:

- Streamlining information delivery and customer service to improve customer support,
- Automating and aggregating purchasing to cut handling and inventory costs,
- Integrating the manufacturing process into the supply chain,
- Bringing distributors into e-commerce initiatives to preserve key customer relationships and support services,
- Overcoming the human issues of trust and confidence—ensuring that proprietary information will not fall into competitors' hands or that customers will not become privy to our inefficient, internal processes.

And yet, e-business will be much more than these functionalities. A small startup based in Seattle, Washington, called 2bot, for instance, has developed a simple yet accurate three-axis milling machine that can be equipped with a laser-cutting head. This device is revolutionary because it can be operated over the Internet and driven like a Windows-based printer using a simple printer driver interface.

While the device and the startup are not quite ready for prime time, the idea is awe-inspiring. A company can issue all its machinists these devices (which retail for less than $5,000) and send them home. Parts can be ordered from these suppliers on demand, quality can be checked in real time, and reorders can be based on speed of delivery and quality. Essentially, a company can now be in the manufacturing business without having any manufacturing facilities! Ultimately, we are seeing a return to the cottage industries and craftsmanship of the Middle Ages, but with the tools of the twenty-first century.

This is but one example of how e-business is dramatically affecting the manufacturing industry, from customer service to new product design (see Figure 1-1). Using the Internet and the myriad tools that support commerce functions, you can find new customers, reduce the costs of managing orders and interacting with a wide range of suppliers and trading partners, and even develop new types of information-based products, such as remote monitoring and control software and other online services.

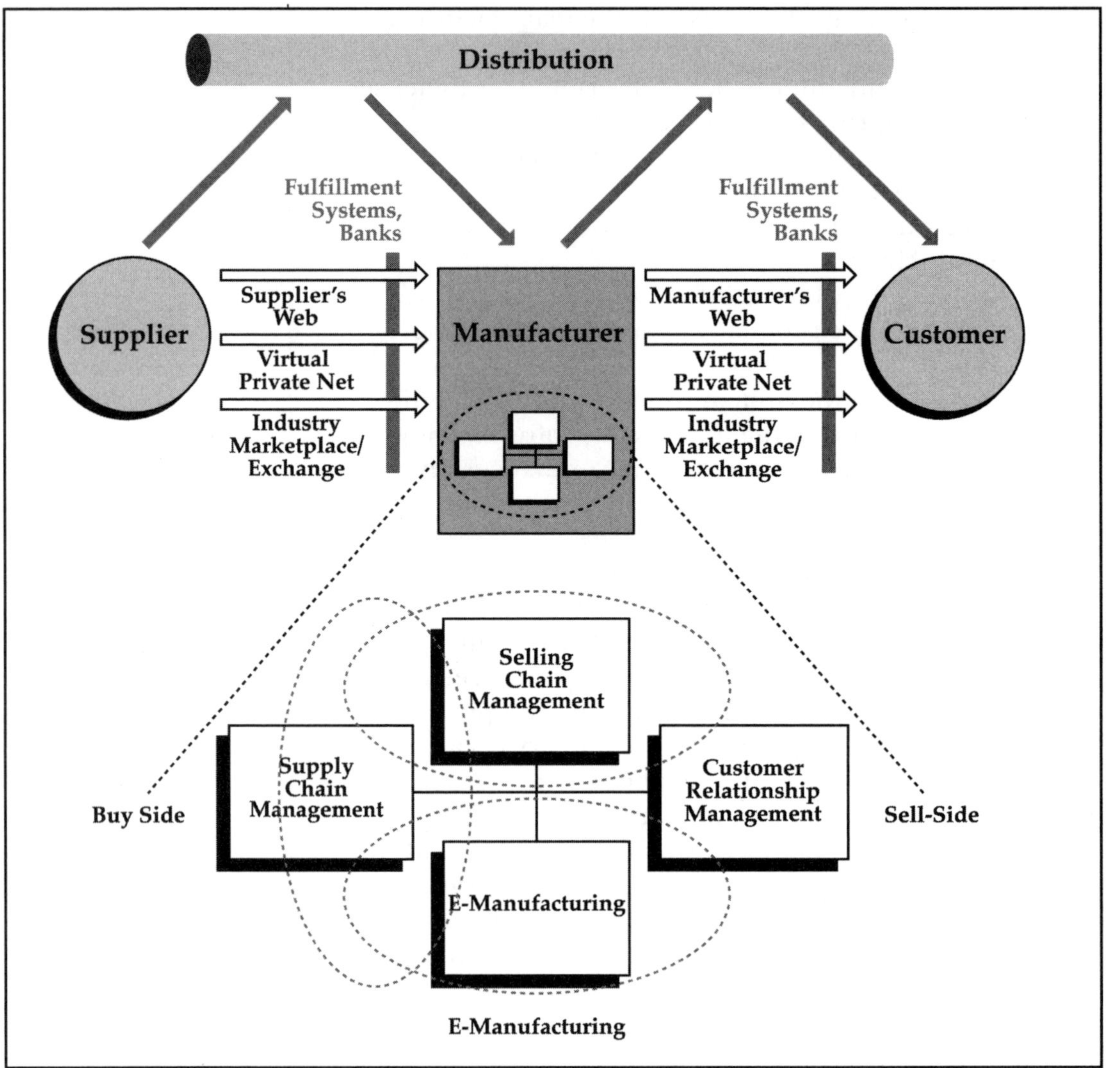

Figure 1-1. Today's e-business environment enables manufacturers to automate and integrate functions as never before—from buy-side to sell-side to manufacturing operations.

Phenomenal Growth

Be assured, e-business is not just a flash in the pan. It is already generating billions of dollars in revenue for many manufacturing companies and is saving millions of dollars for others through online customer support. To provide a little perspective, take a look at the sales numbers for a few select industries in 1996:

- Mail order and home shopping = $80 billion
- Process control equipment = $26 billion
- Video rentals = $15 billion

Back then, e-business was muddling around in the $300 million range, most of which came from connection time charges from the big ISPs and commercial online services. Just four years later, e-business has experienced phenomenal growth, hitting the $400 billion mark in 2000!

It's hard to predict the future, and especially difficult when it comes to the ever-changing Internet. How far will the e-commerce revolution go? Industry analysts are projecting worldwide revenue from online transactions to grow to over $7 trillion by 2004.

Table 1-2. Industry experts may not agree on the exact numbers, but e-commerce is growing at an impressive rate.

(in billion $)	2000	2001	2002	2003	2004
Forrester	$ 604	$1,138	$2,061	$3,694	$6,335
Gartner	$ 403	$ 953	$2,180	$3,950	$7,290
IDC	$ 213	—	—	—	$2,233
Jupiter	$ 311	$ 617	$1,235	$2,128	$3,135
Computer Economics	$3,068	$5,232	$6,815	$9,907	—

Regardless of the exact number, industry pundits (who rarely seem to agree on anything) agree that online direct sales are growing at an impressive pace. According to a recent report released by the Direct Marketing Association (www.dma.org), communications, business services, and industrial equipment top the list of legitimate Internet-based purchases (see Table 1-3). (Sorry, we're not going to cover the porn industry here, even though it has played a leading role in the development of many communications channels over the years.)

Today, most online revenue comes from catalog sales in the technology sector, including such companies as Dell Computer (www.dell.com), Cisco Systems (www.cisco.com), the GE Global eXchange Service (www.gegxs.com), and Avnet (www.avnet.com). By 2002, Internet sales will be divided between catalog sales, online auctions, and contract bidding.

It's no longer possible to simply ignore the results these major companies are generating and the potential for the near future. The bottom line is that we all must do what we can to keep our companies competitive and to provide products and services according to our customers' preferences. In this, the age of networks, that means using the available technology to make your organization a player in the global marketplace.

Regardless of the exact number, industry pundits (who rarely seem to agree on anything) agree that online direct sales are growing at an impressive pace. According to a recent report released by the Direct Marketing Association (www.dma.org), communications, business services, and industrial equipment top the list of legitimate Internet-based purchases.

Table 1-3. Online direct sales are growing.

(in million $)	1999	2000	2001	2005	Compound Annual Growth, 00-05
Communications	$1,281.0	$2,271.6	$3,839.5	$12,621.5	40.7%
Business Services	$1,202.3	$1,886.8	$2,855.1	$6,675.8	26.8%
Industrial Equipment	$1,030.2	$1,583.0	$2,318.5	$4,878.1	25.2%
Instruments	$ 463.5	$ 856.7	$1,487.1	$5,303.4	44.0%
Electrical Equipment	$ 507.7	$ 798.2	$1,203.7	$2,788.7	28.4%
Airlines	$ 414.9	$ 775.6	$1,370.9	$5,327.3	47.0%
Real Estate	$ 375.0	$ 697.3	$1,239.2	$4,651.8	46.2%

Source: Direct Marketing Association, 2000-2001

Because the Internet is worldwide, taking advantage of it can dramatically increase your organization's geographic coverage, giving you a larger market in which to operate. In addition, using an online information center open twenty-four hours a day and seven days a week ("24 x 7") reduces your internal requirement for customer support. This promotes the most important use of the Net—to empower your customers and partners to serve themselves and collaborate! And the tools of e-business are central to making this happen.

EDI and VPNs

Many consider e-business to be an outgrowth of electronic data interchange (EDI). EDI, which has been around since the 1970s, is a standard format for exchanging business data between two companies over a private network (called a Virtual Private Network or VPN).

In the words of Mark Hoffman, president of Commerce One, "EDI is 'dead man walking'."

While EDI allows many manufacturers to tighten up their just-in-time manufacturing and lower supply chain costs, it also has its drawbacks. A store-and-forward technology, EDI is not real-time or interactive, it's expensive, has high transaction costs, and is not widely adopted. As a result, EDI's time is running out. In the words of Mark Hoffman, president of Commerce One, "EDI is 'dead man walking'."

Luckily, e-business is different from EDI. First, it's based on Internet and Web technologies. This means that the communications medium is public—open to anyone with Internet access. Second, because it is based on Net technologies, customers need only ensure that they have one of the common Web browsers to be compatible with virtually all of the online offerings. This makes access interactive, inexpensive, and widely accepted.

So what will happen to EDI? Think Internet, intranets, extranets, and online communities.

Intranets, Extranets, and Online Communities

Intranets are a powerful medium. Internal company networks that are based on Internet technology, intranets empower employees, encourage individual creativity and innovation, and break down barriers between departments by making information available across the company. What more could you ask for from a corporate communications vehicle? Above all, intranets allow your organization to build more efficient core processes in such areas as procurement, supply chain management, product development, sales, and customer support.

Intranets can take several forms. In the static publishing mode, intranets contain employee lists and directories, policies and procedures manuals, organization charts, internal job postings, inter- and intradepartmental reports, product catalogs, newsletters, and training materials. At the next higher level, intranets become a medium for dynamic applications, such as help desks, sales applications and reports, order entry, inventory and production status, and customer service. Once these services are in place, the corporation needs to build the collaboration facilities for the corporate intranet. Such services include e-mail, internal newsgroups and discussion lists, calendars and schedules, video conferencing, group writing or shared documents, data conferencing, and knowledge repositories. Intranets also improve workflow in employee education, product development, procurement, manufacturing, and order fulfillment.

At the highest level of intranet, a secure external gateway is established from portions of the internal network to the outside world. These extranets are the real future of e-business. According to the Gartner Group (www.gartner.com), eighty percent of new e-commerce business opportunities will be by way of extranets. Extranets will replace EDI as the means by which companies manage the procurement process, work closely with suppliers, and provide special information to key customers.

e-Business in Action

So how is e-business actually taking shape? Fruit of the Loom, for example, uses the Web to tie together its network of distributors so they can provide customers with information on their mills and distributors' inventories. To accomplish this, the company implemented an online catalog, which includes a complete product database and ordering information; online customer billing, which routes through accounts payable to handle snail mail billing and payment; and bulletin boards and chat rooms so customers can exchange information with each other.

Computer vendor Gateway 2000, which is already selling over a billion dollars annually online, uses its Web site to provide everything from online ordering to technical support, corporate information, and a humor section with contests and an online magazine.

Before we take a more detailed look at the future, let's first consider the issues that are currently holding us back.

While e-business is absolutely critical to the success of any business that plans on being around through the next decade, it changes the ground rules for everyone. It involves a fundamental shift in internal company business practices and requires riding a not insignificant technology learning curve.

Barriers to e-Business

While e-business is absolutely critical to the success of any business that plans on being around through the next decade, it changes the ground rules for everyone. It involves a fundamental shift in internal company business practices and requires riding a not insignificant technology learning curve.

Even so, the biggest barriers to e-business today come from our customers:

- People don't trust the security of online ordering systems.
- There is no easy way to pay.
- People can't find the information they need online.

A number of secure transaction technologies are in the works, but for the most part e-business is plugging along using the Secure Sockets Layer (SSL) and Transport Layer Security (TLS) network protocols. SSL is primarily an issue for those who wish to pay online with credit cards. However, continuous enhancements to the protocol ensure that security features are steadily improving. For example, biometric identification (e.g., fingerprinting, iris scans) is increasingly being used instead of passwords and client-side payment information management.

As to finding what people need online, companies must be more in tune with their customer base. The ideal Web site provides a high level of personalization, high technical content, straightforward navigation, superior products and services, and a way for customers to interact (to form a sense of community among themselves and with your company).

A Look into the Future

So, where are things headed? Imagine the following happening in the not-too-distant future...

A purchasing agent from an automotive manufacturer needs to place a large order for a digital signal processing integrated circuit (IC) for an on-board global positioning system (GPS) that will be standard equipment in its newest vehicles. The purchasing agent goes to the IC manufacturer's Web site and places the order. An order tracking number is generated by the manufacturer's order entry system and sent to the customer.

At the IC manufacturer's facility, the new order is sent through the enterprise resource planning (ERP) system to the production system, which is connected to each device on the production floor. The production system queries the inventory tracking system and the devices on the plant floor and discovers that the company will need to ramp up volume. The production system revises the forecast and sends the information on the new order, along with the revised forecast, to the production manager for approval.

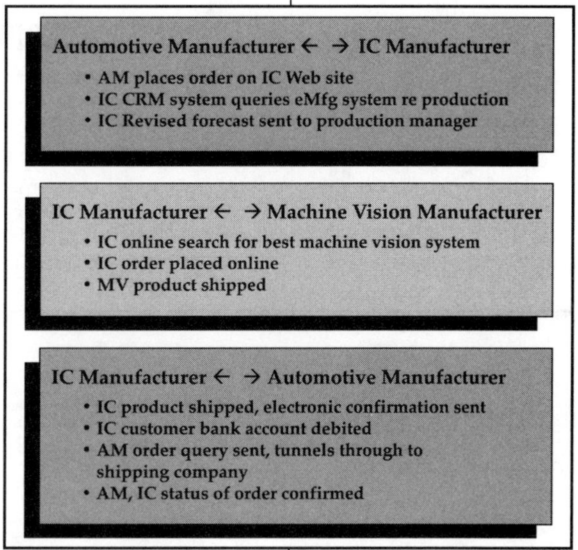

Figure 1-2. In the future, technology will become a non-issue to users—as the Internet and software agents handle most transactions.

The production manager sends two manufacturing engineers off to upgrade one of the production lines. The engineers get online and focus on a machine vision system that is needed to ensure quality at higher speeds. They use a product comparison intelligent agent to ask a few questions via e-mail, then they place an order for the machine vision system, cameras, and additional wave soldering equipment and supplies.

The machine vision equipment is shipped and installed. IC production reaches capacity on schedule. Upon shipment of the ICs, the IC manufacturer's computer automatically debits the bank account of the automotive manufacturer and sends a confirmation that shipment has been made. The day before the first components are due to arrive, the automotive company's order tracking agent queries the IC manufacturer. By way of an embedded extranet, the agent tunnels through to the freight company's computer to track the shipment. If all is well, the purchasing agent is automatically notified. If the shipment is not on schedule, the computer flags the customer, the shipping company, and the vendor, providing them with a status update.

Technology Convergence

What the scenario in the previous section demonstrates is that as Internet technology converges with artificial intelligence, wireless de-vices, and embedded computing, technology will become a non-issue to its users, be they customers or suppliers (see Figure 1-3).

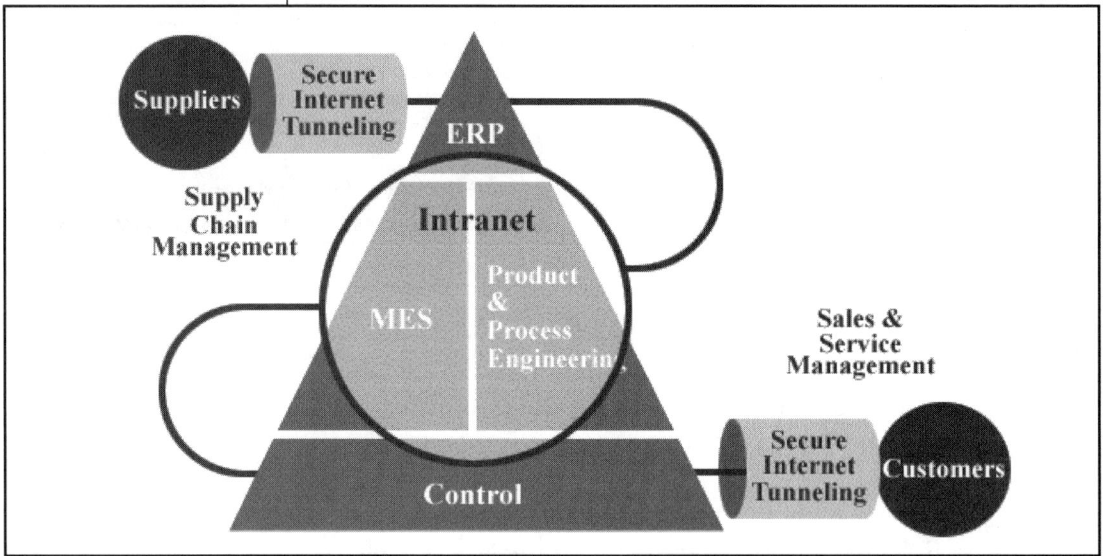

Figure 1-3. As Internet technology converges with artificial intelligence, wireless devices, and embedded computing, technology will become a non-issue to its users, be they customers or suppliers.

When embedded computing technology is combined with the Internet we gain truly intelligent factory floor devices. When wireless technology is combined with the Internet we gain "any-to-any" computing. When intelligent agent software is combined with the Internet we gain automatic, computer-generated interactions that free us from mundane tasks and more closely bind customers and their suppliers. Ultimately, the Internet empowers both people and factory floor equipment.

Basic human psychology will become the primary focus of our service efforts as we work to adjust our processes to meet our customers' increasingly shorter attention spans, their need for immediate gratification, and their insistence on easy-to-use online services. With over one billion Web pages available as of early 2000, our customers have more options than ever before. If we don't provide the level of service and support they want, then there are plenty of other options available to them.

Take something as simple as online book buying. One of our favorites is Amazon.com. But if Amazon.com doesn't have the particular book we're looking for, it's just as easy to buy from Borders.com or Barnesandnoble.com or at least twenty other sites that provide a broad selection and discounted prices. And therein lies the challenge of selling commodity-type products, whether online or off. The reason we like Amazon.com is its additional services. Each week, we receive recommendations about books we might like based on our past buying patterns, and they offer e-mail newsletters designed for niche topics, such as fiction writing or Christian books. It is the value-added services they provide that keep their customers loyal.

The Internet as a "Disruptive Technology"

Fundamentally, the Internet can be looked at as a "disruptive technology"—one that can quickly change the basis of competition. The Internet presents suppliers with new opportunities for developing a unique competitive advantage in their markets. In many cases, we see startups using the Internet to, in effect, sneak up on their competitors through emerging market opportunities.

According to Clayton Christensen, who coined the term *disruptive technology* in his popular book, *The Innovator's Dilemma*, "Disruptive technologies are innovations that result in worse product performance, at least in the near-term. They under perform established products in mainstream markets. But they have other features that a few fringe (and generally new) customers value. Disruptive technologies typically enable new markets to emerge. There is strong evidence showing that companies entering these emerging

The highest-performing companies, Christensen argues, "find it difficult to invest adequate resources in disruptive technologies—lower margin opportunities that their customers don't want—until their customers actually do want them. And by then it is too late."

markets early have significant first-mover advantages over later entrants."[1]

We see this playing out in the rise of B2B (business-to-business) marketplaces and industrial exchanges. e-Steel (www.esteel.com), for instance, provides buyers with increased flexibility, allowing them to "self-serve" as they walk through the products and services available from participating suppliers. In these the early days of such marketplaces, buyers receive increased convenience and lower prices in exchange for the extra time it takes to search the site, find the right product, and place an order.

The highest-performing companies, Christensen argues, "find it difficult to invest adequate resources in disruptive technologies—lower margin opportunities that their customers don't want—until their customers actually do want them. And by then it is too late."[2]

In the case of e-commerce, although only about half of all manufacturers are actually taking orders online, soon this will be the accepted mode for most business transactions.

NAM Survey

In 2000, the National Association of Manufacturers (www.nam.org) conducted the first in a series of surveys about the ways the manufacturing industries are using e-commerce. According to the survey, there is a wide gap between businesses realizing the importance of the Internet and actually putting the Internet to work for them.

According to the survey,[3] sixty-eight percent of the twenty-five hundred respondents said their companies are not currently using electronic commerce for business transactions. Eighty percent of the companies said they have a Web site, but only ten percent say that their current business processes are fully automated. When asked how their companies use e-commerce, they replied as follows:

- 23% use it for new product introductions
- 17% shop for best prices
- 17% use it instead of EDI (electronic data interchange)

1. Christensen, Clayton M., The Innovator's Dilemma. New York: HarperBusiness, 2000, p.xv.
2. Ibid, p.xix-xx.
3. National Association of Manufacturers, www.nam.org/tertiary.asp?TrackID=&CategoryID=439&DocumentID=1955, October, 2000.

- 13% use it for Request for Proposal (RFP)/Request for Quote (RFQ)/Request for Information (RFI)
- 12% buy intermediate materials, parts, subassemblies, or shipping services
- 10% arrange logistics
- 7% buy raw materials
- 5% integrate existing supply chains

In sum, forty-three percent of the respondents said they have an online product catalog that is used for customer service, but only one percent are considered world-class sites. Of the thirty-two percent of manufacturers using the Internet for B2B e-commerce, more than half (52%) said they are using the Web to create new sales channels.

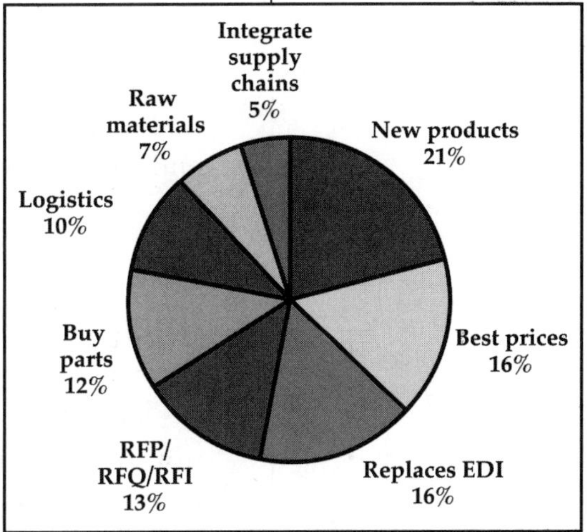

Figure 1-4. According to the National Association of Manufacturers, twenty-eight percent of all manufacturers use their Web sites for e-commerce. The next most popular use is to research new products (21%).

The NAM report concludes that the biggest challenge facing manufacturers is uncertainty about who is in charge of B2B e-commerce efforts. Thirty-five percent said their CEOs or senior management personnel are in charge, twenty-two percent said information systems personnel have the responsibility, and five percent said the responsibility lies with their marketing and sales team. A full ten percent of all respondents admitted that no one is in charge of this important function.

A follow-on survey queried NAM board members about their e-commerce priorities. Cutting costs using online procurement and improving supply chain management are the driving forces behind e-commerce initiatives in industry. According to the survey, the top issues driving e-commerce are the desire to improve customer satisfaction (89%), create new sales channels (70%), reduce order processing and other transaction costs (68%), and increase internal productivity (67%).[4]

Tracking the Best

So what does it take to develop a world-class e-business site? In September 2000, *B2B* magazine ranked the top two hundred B2B Web sites. They analyzed over seven hundred sites, rating them for ease of navigation, presentation of

4. Ibid.

information, quality of information, interactive resources, and e-commerce capabilities. According to their survey, the top ten manufacturing sites were, in ranked order:

- General Electric (www.ge.com)
- Boeing Co. (www.boeing.com)
- Interconnect Devices (www.idinet.com)
- Eastman Chemical (www.eastman.com)
- Aero-Marine Products (www.amp.com)
- Andrew Corp. (www.andrew.com)
- Bethlehem Steel (www.bethsteel.com)
- Miller Multiplex (www.millermultiplex.com)
- Pall Corp. (www.pall.com)
- Sonoco Corp. (www.sonoco.com)

In the high-tech manufacturing sector, top companies included IBM (www.ibm.com), Cisco Systems (www.cisco.com), and Compaq Computer (www.compaq.com).

IT Budgets Drive Innovation

The ability of companies to achieve their goals depends, not surprisingly, to a large extent on the way they spend their IT budgets. Such spending differs widely from industry to industry. How does the manufacturing sector stack up? In a year-end survey of *Information Week* magazine's top five hundred American companies,[5] they found that telecommunications companies led in IT spending but when it came to profits, distribution companies derived more from their e-business operations.

5. Information Week, September 11, 2000, www.informationweek.com/803/splash.htm.

IT spending as a % of projected revenue

Telecommunications	17%
Biotechnology & Pharmaceuticals	5%
Electronics	4%
Manufacturing	3%
Automotive	2%
Chemicals	2%
Distribution	2%
Overall average:	4%

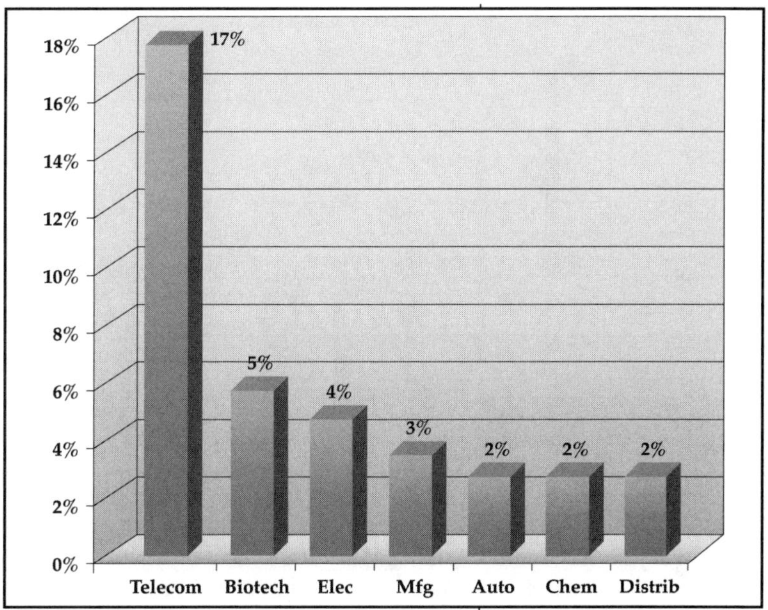

Figure 1-5. Telecommunications tops the list for IT spending as a % of projected revenue, compared with the manufacturing sector's 3%.

Customer service inquiries handled electronically, without intervention

Distribution	20%
Telecommunications	20%
Electronics	19%
Manufacturing	19%
Biotechnology & Pharmaceuticals	16%
Automotive	14%
Chemicals	9%
Overall average:	21%

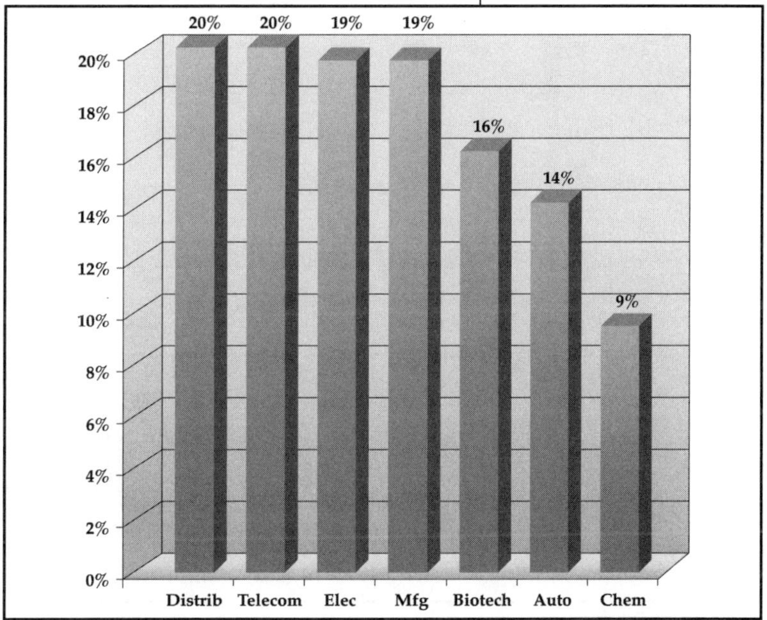

Figure 1-6. A growing percentage of customer service inquiries are handled electronically, without intervention, across a variety of industries.

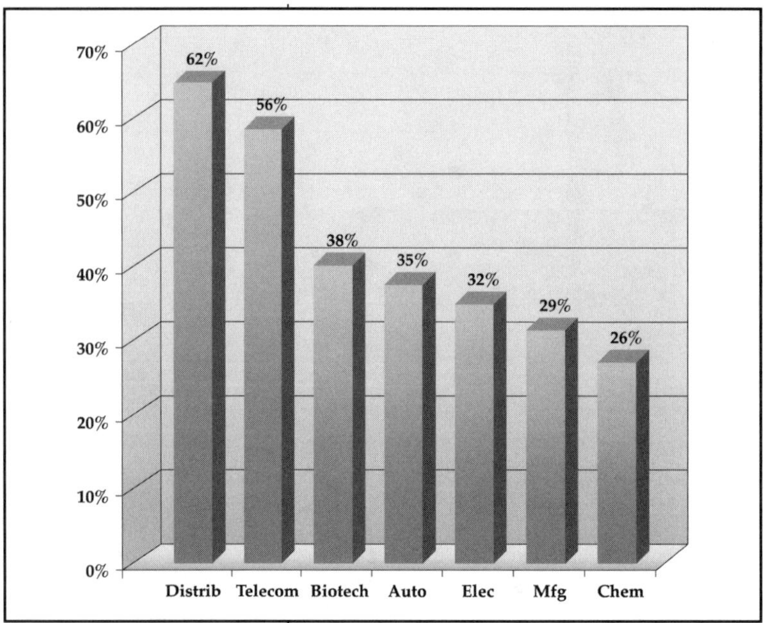

Figure 1-7. Overall, forty-five percent of companies are showing profits from their e-business operations.

Companies showing profits from e-business operations

Distribution	62%
Telecommunications	56%
Biotechnology & Pharmaceuticals	38%
Automotive	35%
Electronics	33%
Manufacturing	29%
Chemicals	26%
Overall average:	45%

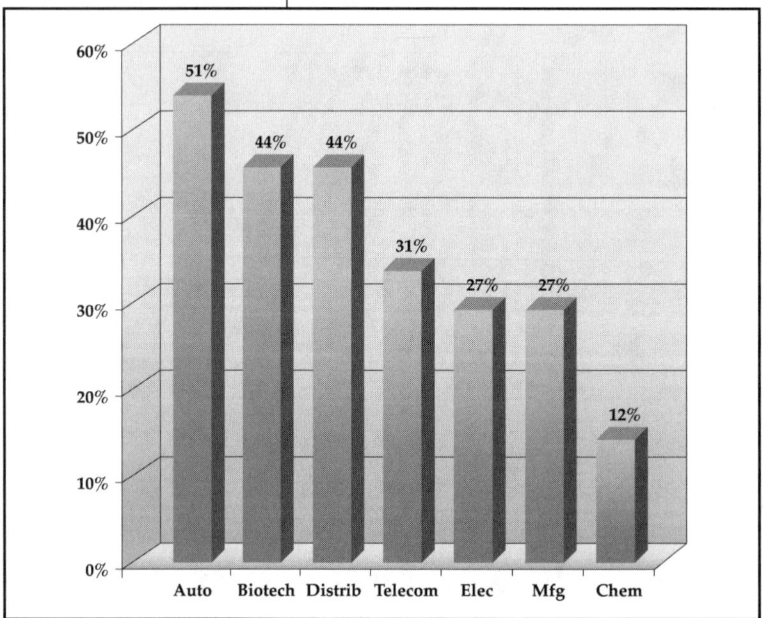

Figure 1-8. The automotive industry is leading the way, integrating more than half their suppliers into electronic supply chains.

% suppliers included in electronic supply chain

Automotive	51%
Biotechnology & Pharmaceuticals	44%
Distribution	44%
Telecommunications	31%
Electronics	27%
Manufacturing	27%
Chemicals	12%
Overall average:	28%

% customers included in supply chain

Automotive	53%
Distribution	51%
Telecommunications	40%
Biotechnology & Pharmaceuticals	27%
Electronics	25%
Manufacturing	25%
Chemicals	19%
Overall average:	34%

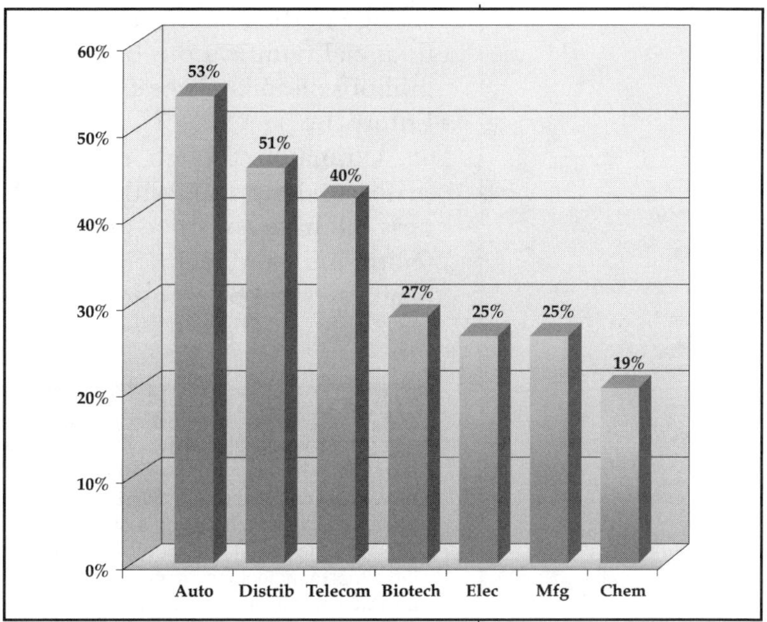

Figure 1-9. Customers are also included in the electronic supply chains, with the automotive industry and distribution channels over fifty percent integrated.

As manufacturing organizations seek to move beyond the "brochureware" stage of Web sites, the goal is to help customers efficiently find and order what they need *and* to eliminate the buffers of inventory between buyer and seller that create waste and cost throughout the supply chain. To rapidly respond to changing marketing conditions, all members of the supply chain must cooperate and coordinate with each other.

In a September 2000 interview with *Information Week* magazine, Tom Orlowski, vice president of information services at the National Association of Manufacturers NAM, observed, "It's no longer about manufacturer competing against manufacturer. It's about supply chain competing against supply chain."[6]

The goal is to help customers efficiently find and order what they need and to eliminate the buffers of inventory between buyer and seller that create waste and cost throughout the supply chain.

It still remains to be seen how the details of e-business implementation will shake out. Technology is moving at a phenomenal pace. Will buying and selling mostly occur through direct links between companies, virtual private networks (VPNs), or online exchanges? We'll cover such issues in more detail in the chapters that follow. For now, suffice it to say that it's anyone's call. Extranets are streamlining selling channels and marketplaces are creating strange bedfellows. Unlikely competitors are teaming up to create new online communities, yet traditional distribution relationships still remain a top concern for most manufacturers. People and processes have always been relatively slow to

6. Information Week, September 11, 2001, www.informationweek.com/803/manufacturing.htm.

change in the manufacturing sectors, primarily because the financial ramifications of an error can quickly run into the millions of dollars (see sidebar, "The Web-enabled Industrial Enterprise").

Ultimately, though, e-business as a special form of enterprise will disappear. Within less than twenty years, all business will have absorbed and adopted the concepts, techniques, and technologies we are discussing here. All businesses will be e-businesses.

The Web-Enabled Industrial Enterprise

There was a fascinating poster hanging in most airports throughout 2000. It was an advertisement for Andersen Consulting (now Accenture). It showed an aerosol can spraying a cloud of letter e's at a conventional enterprise. The point was that simply calling a business an e-business doesn't make it so.

The growth and development of the Internet since 1994 has taken the industrial manufacturing sector by surprise. Not once, but many times, both big and little players in this huge market have been, as the former chairman of Invensys' Action Instruments division, Jim Pinto, puts it, "like deer caught in the headlights."** (See the case study at the end of this chapter for more on the Invensys story.)

Even now, there are companies that still have a single e-mail address; a secretary prints out and distributes the company e-mail in hard copy form for each recipient. There are even significant numbers of companies whose IT departments will not permit free access to the World Wide Web for either purchasing or engineering functions. Fear of viruses, undesired intrusion into company files, or the migration of pornography onto company servers are significantly handicapping what might otherwise be a competitive manufacturer.

A large Canadian oil producer, for example, until quite recently provided a single "Web station" for its engineers to use if they needed to find something on the Web. Currently, once they submit a request and receive permission, everyone at the company may access the Web, but they are limited in what they may do. No FTP access and no streaming media are permitted. This means that web-conferencing and web-based seminars are not permitted. This same company has a twenty-four-hour quarantine on all attachments to e-mail sent to company personnel, except rich text format and Acrobat documents. Though we picked an oil company in this previous example, there are many cases of such thinking in the manufacturing sector.

From Brochureware to Web-based Catalog

In recent discussions with automation and controls engineers, a commonly reported frustration we have encountered is the lack of response to e-mail inquiries. Replies can take days or even weeks to appear. Unfortunately, we're working on "Internet Time" now. That means replies are expected within twenty-four hours...sometimes even sooner.

The Web has gone from being a place to stick up a "business card" Web site to a place for a "product line card" Web site to a slightly more encompassing "Web-based catalog." Most industrial manufacturing companies have finally reached the online catalog stage. But the potential of the Internet is much greater than this, and some companies are just now beginning to realize the power of this tool, not just for marketing but for actually reorganizing their own internal structures.

*** From Pinto's Pointers. For more interesting quips and quotes about the measurement and controls market and trends in technology from industry pundit and futurist, Jim Pinto, see www.jimpinto.com.*

(continued)

The early-stage "business card" Web site was a short, one- or two-page Web document intended to show stockholders and customers that the company was "with it." Even as late as July of 2000, large companies like Schlumberger continued to have only the most basic of sites. Schlumberger's Web site, for instance, was a business card on steroids, designed for stock market analysts and shareholders. A look at www.schlumberger.com (re-designed in December of 2000) shows that you now can find news, the company's stock price, and press releases quickly, even a telephone directory of employees. But several minutes of searching may not disclose the process instrumentation and water meter product line information interested engineers might be looking for. In fact, engineers, like most Net users, will give up trying after only a few minutes, never to return.

If you do manage to find information on the kinds of instruments Schlumberger makes, it will be no more than basic product listings, and a simple technical bulletin. You won't find detailed information, specifications, pricing, and purchasing instructions. In fact, you won't find much more than a line card, which is insufficient to provide engineers and purchasing agents with the critical information they need to make final decisions.

Some manufacturing companies have reached the point where they are duplicating their catalog binders on their Web sites. Some even have interactive tools, such as product configurators, and some even allow direct purchasing. But very few companies have made any attempt to really integrate the Internet into the way they actually do business.

What Manufacturers Aren't Doing

Few companies make it possible for their customers to check order status and inventory information directly online. Even fewer make customers' order history directly available to their customers, and fewer still use the power of Web-enabled communications to improve their ability to provide quality customer service at reduced cost to customers around the world.

Few companies provide their sales reps and distributors with real-time access to product availability and pricing information or real-time backup technical support for field sales. Yet all of these things are not only possible with the current state of Internet technology; they can reduce costs and increase sales at the same time. But few companies have put the technology to work. Yet.

A truly Web-enabled manufacturing company is still to emerge. However, some of the features it will embody have become clear. It will be:

- Flat and heterarchical**
- Organized on the principle of "no-drag" information transfer between all parts of the company
- Organized on the principle of "no-drag" information transfer between the company and the customer
- Customer-oriented and customer-driven
- Nimble and efficient
- Knowledge-based
- Run by empowered employees at every level

*** Heterarchical is the opposite of hierarchical, but not non-hierarchical. A heterarchy is an organization that has many different organizational dynamics, one that is inherently organized pragmatically around what works. For more on heterarchies, see Funky Business: Talent Makes Capital Dance by Riddersråle and Nordström, published by ft.com, 2000.*

Some manufacturing companies have reached the point where they are duplicating their catalog binders on their Web sites. Some even have interactive tools, such as product configurators, and some even allow direct purchasing. But very few companies have made any attempt to really integrate the Internet into the way they actually do business.

CASE HISTORY CASE HISTORY CASE HISTORY

Invensys: iastore.com

To survive, even thrive, companies learn that when customers speak, they must listen. And in early 1995, when Foxboro's customers started talking about, even insisting on, electronic order processing, the automation supplier listened. Foxboro's Fortune 500 customers were pushing for self-service—to make their purchasing methods more productive and to drive down costs. Each looked to the new hyperlinked environment of the Web to streamline relationships with their suppliers, like Foxboro.

So Foxboro embarked on a major study. The company's Pricing Policy Group, initially an IT-driven group with some marketing participation, asked the company's six largest customers about their purchasing needs and long-term goals. The results of the study evolved into a requirements document for webstore.foxboro.com in 1996, one of the first Web-based spare parts ordering systems in the automation industry.

Foxboro initially contracted with an outside software development house to convert its requirements documentation into a functioning Internet site. Although the group it chose ended up falling to the wayside after a buyout, they did provide the Foxboro IT group with a defined systems architecture and the initial hardware and software structure. From there, Foxboro programmers took it upon themselves to complete the first, rudimentary storefront.

By the end of 1996, webstore.foxboro.com was seeing a fair amount of activity—ten thousand user sessions per month. Unfortunately, they only received about ten orders per month as a result of that activity since most internal and external people were using the site simply to look up part numbers and pricing. Virtually all orders were still placed via phone or fax.

Over the years, webstore.foxboro.com took on a life of its own. Initially, a Foxboro-only site, in 1997 a dozen sister companies' products were added from the Siebe Group of companies (at that time, Siebe was Foxboro's parent company; now it is an Invensys company). Six months later, in 1998, Foxboro converted an extensive series of desktop configurators and quote engines (which had initially been developed long before the arrival of the Web) into a powerful new set of tools for quick and easy ordering. Then, in 1999, iastore.com launched an online order tracking system.

As the store evolved through these stages and more functionality was added, activity grew. As of 2000, Foxboro was seeing more than fifty thousand user sessions per month and millions of dollars in online orders. With the arrival of the online order status function, the bulk of Foxboro's service calls disappeared. Before 1999, the iastore companies received between 650 and 750 calls per week simply inquiring about order status. After the completion of the order tracking function, the number dropped to fifty per week.

From HTML to XML

The bulk of iastore.com consists of HTML coding and database-driven content. While the fifty Invensys companies that are participating in the store use more than a dozen different ERP systems, Foxboro decided to standardize the bulk of the store's functions to run on SAP, its internal ERP system. Other sister companies use Marcam or Baan or PeopleSoft systems. To handle this integration, the company decided to employ the eXtensible Markup Language (XML) format. XML serves as the translation interface for seamless information transfer between the storefront and each company's ERP system.

Although XML is an industry standard coding language, there's a lot that isn't standard about it yet. Several different groups are haggling amongst themselves about various formatting issues. So Foxboro decided to start with what was available and then customize its code. It started with the generic XML data type (based on the work being done at OBI and Biztalk), then customized it to handle the specifics of its various ERP systems.

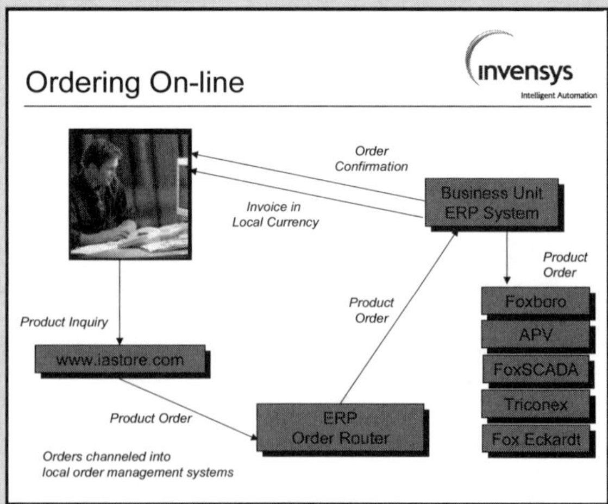

In addition to the standard Web interface, today iastore.com uses more than fifty extranet connections to its major customers. XML plays a role here as well since customers are fairly evenly split between Ariba and Commerce One marketplace programs.

Challenges Faced

Many automation companies find that one of the biggest challenges they face when implementing e-commerce is that they aren't able to take orders in an electronic format. Foxboro's director of e-business, Karen Turner, credits the company's foresight to its ability to excel in this area since, early on, Invensys pushed electronic ordering in its companies.

Foxboro also faced a significant hurdle when the company converted its ERP system from Cardinal to SAP a mere eighteen months into the life of Webstore.foxboro.com. While there was little external evidence of the commotion behind the scenes, this was a major transition. The transition took almost two years to migrate through over fifty business units. The support required to make possible different data feeds to different business units represented a significant maintenance effort.

Finally, as with most automation firms, Foxboro also had to deal with fear—from reps, from some purchasing agents, and even from internal Foxboro people—that the Internet would make them redundant. To counter this fear and to educate all involved parties in the ways the Internet could make them more effective in their jobs the company ran a series of seminars for employees. In addition, specific promotions offering discounts to customers who used the store encouraged acceptance.

Internally, Invensys's worldwide intranet keeps employees learning and connecting. Online functions include technical assistance, marketing information, technical documentation, testimonials, corporate presentations, and project discussion groups.

The Next Step

A truly successful commerce site provides detailed product information, easy-to-use product configuration engines, and secure transaction processing. By the end of 2000, iastore.com had added a major new feature—an online catalog, complete with specs, technical documentation, maintenance information, and more. Now the store offers a more complete purchasing experience for its customers, both engineers and purchasing agents.

The future holds interesting possibilities. Director of e-business Turner talks of Foxboro's new global registration system in which any party at any location can log onto the company's Web site and gain access to any part of the internal or external knowledge base, as needed (and with appropriate clearance). A new, patent-pending process will allow engineers, purchasing agents, and salespeople to share sensitive information electronically in a safe, secure environment. Last, but not least, Turner has begun implementation of a series of segmented storefronts that more directly address specific customer needs. All "powered by iastore.com," these storefronts are broken down into such front ends as Measurement Express and APVExpress. While this makes life easier for the small to medium-size accounts, the larger, Fortune 500 customers tend to purchase products across multiple divisions. Providing an easily segmented directory of all Invensys automation products is a challenge that still remains to be solved.

CHAPTER 2

e-Business Basics: Tools of the Trade

The Internet, that now ubiquitous "network of networks" connecting hundreds of millions of people globally, was originally launched by the U.S. Department of Defense as ARPANET back in the 1960s. At that time—the heyday of the Cold War—scientists were trying to build a robust distributed network that could transfer messages over long distances even if random sections of the network went down because of catastrophic attacks. The idea was to be able to create an electronic message, break it up into multiple segments, transfer the individual segments over different parts of the network, and then reassemble the components at their final destination into one, readable piece.

Thus was born TCP/IP (Transport Control Protocol/ Internet Protocol), the protocol spoken by the host computers, also known as servers, of the Internet. TCP/IP is a family of protocols that define how certain applications are to be accomplished on the Internet, including making connections, transferring files, and messaging.

In the early days, TCP/IP was the common hardware language that tied together hundreds of military and academic organizations. Other languages tried to establish a foothold, including ISO/OSI X.25-based packet switching in Europe and BITNET in the university sector. The commercial arena also saw several proprietary protocols emerge, such as CompuServe, America Online (AOL), and AppleLink (a leading-edge e-mail network developed by AOL for Apple Computer, who sold it back to AOL once TCP/IP became the de facto standard). Ultimately, TCP/IP emerged as the protocol that binds the world together through FTP, e-mail, and Web services (see Figure 2-1).

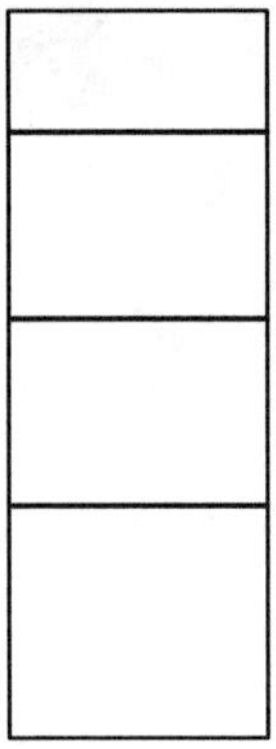

Application or process layer, an application protocol such as e-mail.

Transport layer, where the TCP protocol actually runs, connects the destination device with the data and provides error control for lost or damaged information packets.

Internet layer, provides the addressing needed to allow routers to forward packets of information across local networks.

Network interface, manages the exchange of data between physical devices and the network, such as computers, telephones, telephone networks, and satellite circuits.

Figure 2-1. TCP/IP divides its networking functions into four layers roughly corresponding to the ISO's Open Systems Interconnection (OSI) standard.

The Hardware Infrastructure

The Internet may sound like a simple enough concept, but it involves a complex array of hardware, software, and the data packets that are transported from one location to another. Its major components include the backbone networks, the local connection points, and the customer premises equipment (CPE, that is, the network equipment located at an individual's or customer's site).

Once controlled by the National Science Foundation, NSFNET was the original backbone network over which all Internet traffic had to travel. To meet expanding demand, the U.S. government sold NSFNET to Advanced Network & Services, a partnership between IBM and MCI, that expanded capacity from T1 (1.544 Mbps) to T3 (45 Mbps). Today, the backbone is owned by several commercial firms, all of whom are racing to keep pace with the Internet's relentless growth.

Local connections to backbone networks are provided by Internet Service Providers (ISPs) or commercial online services. ISPs are regional or national organizations that have purchased high-speed connections to backbone providers. ISPs then sell a piece of their bandwidth to individuals or businesses through various service packages, which can include Web hosting, e-mail, and e-commerce capabilities.

Commercial online services, such as AOL and Microsoft's MSN, are closed, proprietary networks that provide several special services, such as news, games, and member-to-member chat, in addition to various degrees of external access to the Internet for e-mail, Web browsing, and most recently, chats between nonmembers.

When choosing your Internet access provider, you will find that most provide the same level of functionality for individual customers. Commercial services, such as AOL, tend to be easier to install and so appeal to both individuals and small companies. Firms who are serious about providing Internet access to many of their employees should work with an ISP who has a reputation for high uptime and fast customer service.

Network Equipment

The Internet interfaces to the enterprise through customer premises equipment (CPE)—that is, the privately owned routers, gateways, switches, and associated wiring and software that connect to computer servers. Switching is a major technology component of the Internet. The Internet relies on fast packet switching to move bundles of data (packets) through a network at very high speeds.

One of the biggest advances in the communications industry is optical networking components. Traditional optical-switching methods meant converting signals to electrons, then using integrated circuits to convert the signals back to photons. By using a MEMS[1]-based all-optical switch, today's optical routers can handle exceptionally large amounts of data, up to a petabit, that is, 1,000 terabits.

Routers are internetworking devices that intelligently connect local area networks (LANs) and backbone wide area networks (WANs). Routers allow companies to segment their networks so if a problem develops on one it won't bring down a department that is running on another. Hubs are the wiring centers for large LANs. They are used to diagnose line failures, manage traffic flow, measure traffic, and simplify the process of reconfiguring large LANs.

Not everyone connects to the Internet through a direct T1 or T3 line. Many individuals and small companies rely on a combination of modems and standard phone lines.

Currently, the fastest dial-up connections run at 56 kbps. Shared, "always-on" connections include ISDN, xDSL, and cable modems. ISDN (for Integrated Services Data Network) operates at roughly 128 kbps, while the various types of DSL (Digital Subscriber Line) vary between 256 kbps and 600 kbps. Cable modem connections run at between approximately 600 kbps and 1,200 kbps, but they can run slower, depending on network congestion and the number of other users on the cable. For example, the computer on which this is being typed is connected to the Internet via cable modem at 444.83 kbps (clocked using McAfee's Internet Speedometer software).

1. MicroElectroMechanical Systems (MEMS).

Servers

The final element in any e-commerce system is the "server," which is a computer and its associated software designed to manage network resources. Servers, which come in many different flavors, receive requests from a third party and respond to those requests by delivering a particular type of information. For example, database servers process database queries, network servers manage network traffic, and file servers store files for any user on the network. Most servers are dedicated to a single task.

Requests are sent to a server via a "client," such as when a Web browser (a kind of client) interfaces to a Web server to retrieve Web pages. Servers began life as fairly simple devices that simply responded to requests without processing the retrieved information in any way. These early "Web servers" led to the development of "application servers." These are more intelligent devices that process information and interact with the client in such a way that give and take between the computer and user becomes possible.

Proxy servers cache frequently accessed data, which means they can deliver information to the browser more quickly and preserve more bandwidth for accessing new Web information. In addition, they have become a popular way to bridge local networks, LANs, or intranets to an external network such as the Internet.

The workhorse of the Internet is the FTP (File Transfer Protocol) server. An "on-demand server" that moves files around the Internet, the FTP server allows clients to upload files to or download files from a computer. Today's FTP servers provide their administrators with increased control over who can connect and share files, what types of files can be shared, and where those files are placed.

Client/Server Architecture versus Point-to-Point

Point-to-point communication is one of the oldest ways of connecting computers and devices for exchanging data. Under this arrangement, each computer on the network has equal responsibilities. These are fairly simple networks, although they can't always provide a sufficient level of performance under heavy loads.

For that reason, most computers are connected to the Internet using a client/server architecture. In this arrangement, each computer on the network is either a server (as described above) or a client, that is, a PC or workstation that runs software applications. Clients rely on servers for accessing files and for processing power.

Point-to-point (P2P) is making a comeback, however, as companies like Napster use this architecture to permit music to be shared from computer to computer. Although Napster may have run afoul of the copyright laws, new uses for distributed point-to-point computing are being found daily. The SETI-at-home project showed that is was possible to create a "supercomputer" using the pooled dead time (i.e., when your computer is on but inactive) on thousands of computers worldwide. Companies are now looking at the possibility of using point-to-point technology to distribute calculation-intensive projects across their networks and the Internet. In addition, P2P theorists have noted that there is less overhead required with an Internet-enabled P2P network, which means that the shifting requirements of sales and service organizations may be better served with P2P architectures than traditional client/server models.

Point-to-point (P2P) is making a comeback, however, as companies like Napster use this architecture to permit music to be shared from computer to computer. Although Napster may have run afoul of the copyright laws, new uses for distributed point-to-point computing are being found daily.

The Software Infrastructure: Internet Protocols

Contrary to popular opinion, the Internet wasn't an overnight success. It took many years to develop. What did come close to being an overnight sensation was the World Wide Web, a client/server software architecture developed by Tim Berners-Lee of MIT (but then at CERN, the European Laboratory for Particle Physics). Introduced to the world in 1994, the World Wide Web uses a variety of standard protocols to organize and present information and services across the Internet.

As Brian Thomas noted in his landmark book, *The World Wide Web for Scientists & Engineers* (1998), the Web was designed to "provide a common (simple) protocol for requesting human readable information stored at a remote system, using networks."[2] The overarching objective was to give scientists a way to exchange data—whether text, graphics, figures, or databases—using a concept known as hypertext (described in the next section). From this original manifestation as a test bed of technical professionals, the Web has exploded into a communications phenomenon.

As previously mentioned, the Web uses the client/server model of computer interaction. HTTP (Hypertext Transfer

2. Thomas, Brian, *The World Wide Web for Scientists & Engineers.* Bellingham, WA: SPIE Press, 1998, p.43.

Protocol) transmits information between Web servers and browsers. Assuming that you have the right software add-ons for nonstandard Web page functions (such as Flash or Shockwave), all Web browsers can communicate with all Web servers. Talk about open source in practice!

Hypertext, HTML, and HTTP

The dawn of the new millennium has brought a plethora of products and services that integrate voice, video, and data over the Internet. Many companies have begun providing ultralow-cost phone and fax services over the Net (no long distance charges!). New network equipment has taken this a step further by integrating the circuit-switching capabilities of voice networks with the packet-switching capabilities of data networks. The goal is to enable integrated communications with customers and among employees, as in the twenty-first-century call center, which provides voice and video connections to its Web site visitors.

The Web relies on hypertext, which is a system of point-and-click links between pieces of information that enables a user to jump from one point on the Internet to another without the traditional reliance on a hierarchical information storage method (see Figure 2-2). For instance, you can read a product data sheet online by starting with the overview, progressing to features and benefits, next analyzing the detailed product specs, and finally ending up at the ordering guide. With hypertext, you have the flexibility to peruse the document according to your needs (or whims, whichever the case may be!). You could start at the specs first, proceed to the ordering guide next, and then move to the related technical notes. The beauty of hypertext is its ability to produce large, complex, richly connected, and cross-referenced bodies of information.

HTML (Hypertext Markup Language) is the language used to format hypertext information on the Web. Hypertext developed out of an experiment with digital photocompositors for typesetting applications and a type of coding used in those devices called SGML (Standard Generalized Markup Language). Early word processing programs, such as WordStar and WordPerfect, used SGML tags directly. For example, if you wanted to print in italics in WordPerfect, you would type <italics>word</italics> and the word between the tags would appear in italic type.

HTML uses a similar system of text tags to mark text so it is displayed in a particular way on the computer. The Web site designer can mark text with standard formatting codes, such as italics, boldface, or table format. A Web browser—the client software program that is capable of reading these codes—then converts the HTML formatting into a displayable page of information. As a simple example, you can boldface text by typing <bold>word</bold>.

XML (for eXtensible Markup Language) is a new and expanded use of markup language under development. While HTML can tag text for visual display, XML can tag text with much richer data, enabling it to be used for e-commerce and other information-sharing activities.

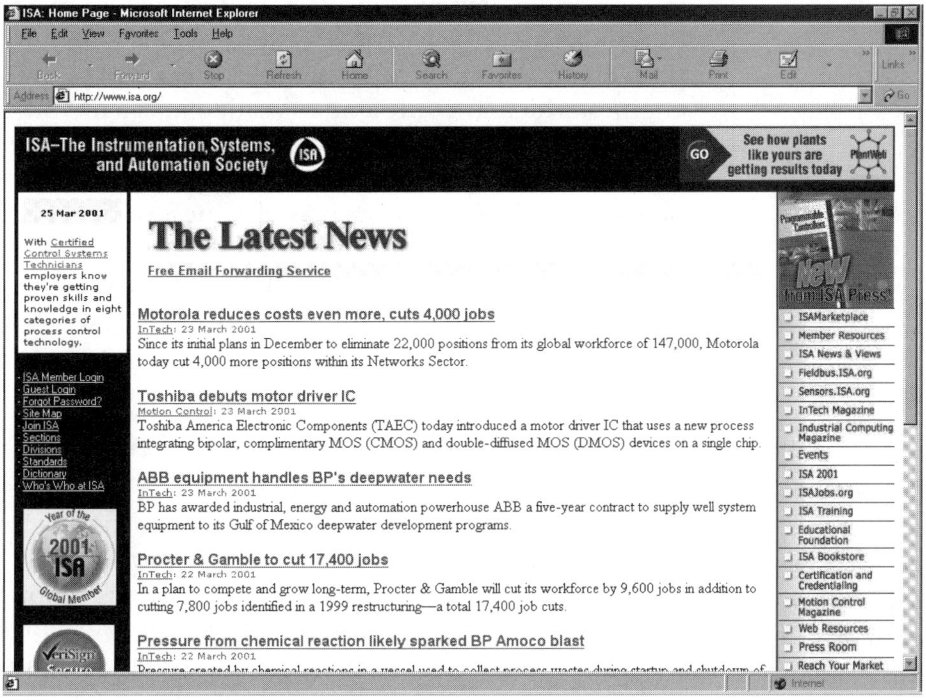

Figure 2-2. News Page on Intech Magazine's Web Site at www.isa.org.

Navigating the Internet

All servers on the Internet are assigned a unique name. Known as a Uniform Resource Locator, URLs are ways of specifying the location of information online. The location can be as broad as a company name or as specific as an image in a document. The URL provides the link between HTML files and HTTP servers and is the way hypertext links are named and can connect two to different places on the Internet.

For example, *telesian.com* is the domain name for the server through which this computer connects to the Internet. The company, Telesian Technology, is running Web server software, which lets you access the corporate Web site by typing the following address into your browser:

http://www.telesian.com

There are two types of browsers: graphical browsers, including Netscape and Microsoft Internet Explorer, and text-based browsers, such as Lynx. Most people use graphical browsers, but a small segment of the population prefers the faster-loading, text-only displays. Unfortunately, most Web sites today make heavy use of graphics. This is a twofold problem in that users face slow load times and search engines, which are only able to read text, are unable to index sites that place most of their content in graphic objects. In addition, the use of heavy graphics disadvantage people with disabilities, such as blindness or dyslexia, who use text-only browsers and screen-reading technologies. Many industrial enterprises employ such reading-impaired people, including Ford Motor Credit's CEO and chairman, Donald Winkler.

Making the Web Work

Today's industrial enterprise must present technical information to its customers, distributors, and suppliers in a logical, easy-to-use format.

Ultimately, the Internet is a networking protocol, TCP/IP, that has a suite of application protocols, such as HTTP, FTP, and SMTP, running on top of it. Everything is standardized, which makes it easy and efficient to access information.

In general, HTML coding is a straightforward process and can be learned in a few hours. But today's corporate Web sites are much more complicated than a few pages of text with some HTML tags scattered throughout. Today's industrial enterprise must present technical information to its customers, distributors, and suppliers in a logical, easy-to-use format. Considerations must be given to the following aspects of the site:

1. **Look-and-feel**—Don't forget that your customers' expectations rise as they regularly visit increasingly sophisticated Web sites, from the family home page to the corporate sites for Hewlett-Packard and IBM.
2. **Information flow**—The rule of thumb is that a user should be able to get to any piece of information in three mouse clicks.
3. **Technical content**—Your customers are at your Web site to get information about your products; don't restrict their access to nonproprietary information unless you're looking to thwart sales!
4. **Interaction**—The best way to build customer loyalty is to provide detailed technical content and a way for the customer to interact with your company via the Web. This can be as simple as two-way listservs or as involved as Flash-based product demos and Web-based videoconferencing.

To truly put the Web to work for you as a competitive weapon, you must also move toward dynamically generated content; the ability to handle online payment transactions; the capacity to interface with internal databases that track inventory, order status, and shipping; and an automated communications network connecting external suppliers and major customers.

Even these aspects merely cover the basics of developing an effective Web presence. To truly put the Web to work for you as a competitive weapon, you must also move toward dynamically generated content; the ability to handle online payment transactions; the capacity to interface with internal databases that track inventory, order status, and shipping; and an automated communications network connecting external suppliers and major customers. These days, undertaking such endeavors moves you into the XML space.

XML

XML is essentially a programming language that describes and exchanges data in an open format. Like HTML, XML is a subset of SGML, the de facto Internet standard for presenting tagged information online. XML is not tied to a particular browser. It addresses content issues and gives meaning to data, in contrast to HTML, which simply handles the presentation of data.

According to Kevin Dick in his eminently readable book *XML: A Manager's Guide,* XML is the tool of choice for several Web-based applications[3]:

1. Synthesizing data from different databases into different packages. This will depend on the needs of each user; some, for example, may want to provide suppliers with access to more detailed corporate information—forecasts, repair orders, returns, and the like.

2. Providing employees with a coherent view of the data. For example, XML could facilitate the design of a Web-based customer service database that integrates order status, shipment details, account information, product configuration, and repair information.

3. Relating different files and formats to each other and to another file. For example, in a Web-based presentation, XML would ultimately enable you to create a seamless "information package" that makes all the elements useful.

4. Providing a general-purpose approach to spontaneous information exchange. For example, online product search and comparison features are easier to design with XML.

What does all this mean? Well, think about interfacing the information in your online product configurator with

Mobile computing is coming to the manufacturing enterprise, although not as quickly as such industries as the financial markets. Within manufacturing, inventory management, equipment monitoring, and employee access to production data are prime candidates for the efficiencies promised by the swing to wireless. New applications for the Palm OS and Windows CE (aka Pocket PC) operating systems enable manufacturing personnel to use handheld computers as programming and data readout devices. The addition of handheld capabilities to wireless phones will make possible seamless access to production data over the Internet or telephone. Such flexible, real-time access to information about workflow and machine utilization will benefit line managers as well as plant floor personnel.

3. Dick, Kevin, *XML: A Manager's Guide. Reading,* MA: Addison-Wesley, 2000.

```
<HTML><HEAD>
<TITLE>Telesian:E-Business Solutions</TITLE>
<META NAME="KEYWORDS" CONTENT="industrial marketing, web development, web design, web site design,
web page design, web design services, public relations, press releases, internet marketing, online marketing, direct
marketing, e-commerce, electronic commerce, strategic marketing, marketing plans, marketing programs,
marketing consultants, marketing services">
<META NAME="DESCRIPTION" CONTENT="Industrial Marketing, Web Development, Public Relations, &
Electronic Commerce Services by Telesian Technology Inc.">
</HEAD>
<BODY BACKGROUND="" BGCOLOR="#FFFFFF" TEXT="#000000" LINK="#0088FF" VLINK="#AA0000">
<TABLE BORDER=0 CELLSPACING="0" CELLPADDING="0" WIDTH="100%">
 <TR>
  <TD BGCOLOR="#000000"> <!– #BeginLibraryItem "/Library/header.lbi" –><TABLE BORDER=0
CELLSPACING="0" CELLPADDING="0" align="center">
    <TR>

<TABLE BORDER=0 CELLSPACING="5" CELLPADDING="2">
 <TR VALIGN=Top><!–<TD><IMG SRC="120pixels.gif" WIDTH="120" HEIGHT="1"></TD>–>
  <TD COLSPAN="2">
   <p align="center"><a href="email_page/frm_email.cfm"><font face="Verdana" size="1">Email
    this Page</font></a></p>
   <p><font face="Verdana" size="3"><b>Telesian:E-Business Solutions</b></font></p>
   <p><font face="Verdana, Arial, Helvetica, sans-serif" size="2">Having worked
    with engineers and scientists for over 20 years, Telesian understands
    the issues facing the manufacturing community and its distributors. Today,
    integrating e-business into the manufacturing enterprise is all about:</font></p>
   <ul>

...

    <li><a href="Web_Dev_IB.html"><font face="Verdana" size="2">Telesian:E-Business
     Internet capabilities</font></a> </li>
    <li><a href="Web_Dev_Best.html"><font face="Verdana" size="2">The best
     sites focus on the customer</font></a>
    <li><a href="Web_Dev_Well.html"><font face="Verdana" size="2">What a well-designed
     web site can do fo you</font></a> <font face="Verdana" size="2"> </font>
   </ul>
   <P> <font face="Verdana" size="2"> </font><font face="Verdana" size="2"></font><!– #BeginLibraryItem
"/Library/text_footer.lbi" –><font face="Verdana" size="2">Shari Worthington, President of Telesian Technology
Inc., is available for public or private <a href="seminar_services.html"> speaking
engagements</a> on E-Business in the Manufacturing Markets, Electronic Commerce,
Internet  Marketing, and High-Tech Marketing. She is currently authoring
a book for ISA on &quot;E-Business in the Manufacturing Industries.&quot; For
more information, e-mail <a href="mailto:lindac@telesian.com">lindac@telesian.com</a>.</font><!--
#EndLibraryItem –>
  </TD>
 </TR></TABLE>
```

Figure 2-3. HTML Sample Code.

the configuration engine run by your major distributor. Essentially, you have two different databases of information residing on different servers. Your distributor could opt to simply link his customers to the product configurator on your Web site, but then he would lose the ability to integrate the part the customer is interested in with his own order entry or inventory management system. Both your and your distributor's configurator database hold such information as part number, product description, optional features, pricing, location in the stock room, and so on. But each company has created a database best suited to its own internal needs. And the chances of your inventory bin locations being the same as your distributor's are virtually nil.

Then there's the issue of upkeep. While you may be unfailingly punctual about updating the product information in your configurator each morning, the distributor may not have the kind of resources it takes to update all of its vendors' information on a daily basis. Besides, why should you and he duplicate information? What's needed is a way to integrate the information from your database with the information your distributor needs. This is the role of XML.

XML

Many people have high hopes for XML. Some claim that this new technology could solve the "islands-of-computing" problem for businesses that have been reluctant to invest heavily in ERP, customer relationship management (CRM), and supply chain management (SCM).

Zona Research recently surveyed two hundred companies with annual revenues of over $1 million and anywhere from twenty-five to twenty-five hundred employees. These companies' reported that the main reason they are working with XML is to create a way to permit enterprise-wide searching (fifty-three percent of respondents) and to shorten application development cycles and enable new operations (both at fifty-two percent of respondents). Surprisingly, given all the recent hype over B2B, the capacity to participate in a B2B exchange was only the fourth most popular reason for delving into XML.

Thus, XML is a way to share information between loosely coupled systems. It's less expensive and less complex to implement that traditional EDI, but XML is not without its own problems. For instance, XML is a memory hog. It produces large data packets that require more computing power than programming languages like C or C++.

Metadata and Shared Context

To truly understand XML, you must first understand the concepts of "metadata" and "shared context." IBM pioneered these ideas several decades ago when they were looking for a way to publish the same content in different ways. These concepts became the SGML standard in 1986.

Metadata is a formal way to describe a piece of information. In other words, metadata is data about data. For instance, "laptop" is the metadata term for my Sony Vaio 505, my trusty book-writing companion. Shared context is a formal description of the rules that metadata must follow. If we're talking about sharing product configurator information, again, the metadata would include "product name," "temperature specs," "SKU," and "Unit Price." The shared context rules could be that "Unit Price" is a number with ten digits, two decimal places, and is always in U.S. dollars.

Kevin Dick states that "The true power of XML is to improve business processes that put multiple documents in the same format. Then a single piece of software can process them all and a single screen layout can display them all. If the format is publicly available, then anyone can generate a document that can be processed by the software or displayed in the screen layout."[4] While XML is convenient in many ways, it is no substitute for such middleware as application servers. You must still provide the technology to move the XML documents from application to application.

The Basics of XML

The XML concept is simple (see Figure 2-4). Programmers add metadata by using "tags" that are similar to those of HTML. To indicate that the Sony Vaio is a laptop, for instance, you would write "<laptop>Sony Vaio</laptop>." Shared context is added to a site through document type definitions (DTDs). A DTD is a set of declarations that specifies the order, structure, and attributes of the tags for a document. A document references a particular DTD either inline (that is, internally within the document) or with an Internet URL. Therefore, any part can spontaneously access the DTD, interpret its rules, and process the document. A parser then reads the XML document and extracts the information so it can be read by another program.

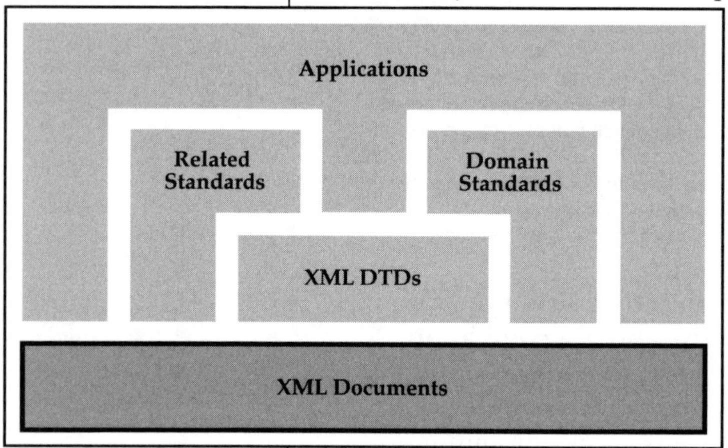

Figure 2-4. The XML Matrix (Dick, Kevin, XML: A Manager's Guide, p. 20, fig. 2.1. © 2000 Addison Wesley Longman, Inc. Reprinted by permission of Pearson Education, Inc.).

Some applications only use XML documents, while others also add DTDs. DTDs enable companies to ensure that they are using the same data format. This leads us to one of the biggest problems with XML—the lack of industry standards. It is not uncommon to hear today's XML described as a hodgepodge of proprietary and complex native interfaces. Basically, Company A cannot exchange XML data with Company B unless they agree on a common set of DTDs. Accomplishing such agreement without a set of industry standards is what turns an XML programming project into a nontrivial expense.

4. Ibid, p.35-36.

XML Standards

XML can best be used by specific industries that are able to agree on a common metadata format. This is not a minor issue. Just consider the differences in the way products are described in the instrumentation and pharmaceutical industries. While instrumentation vendors frequently sell their products to pharmaceutical manufacturers, each company's metadata definitions are different, which leads to major complications when integrating data processing structures.

To solve this problem, industry standards are now beginning to emerge. One of the first to appear was RosettaNet (www.rosettanet.org), a consortium of over three hundred computer manufacturing vendors that are working to define an XML standard for exchanging business data within their industry. The effort, which involves manufacturers, distributors, and suppliers, got underway because computer vendors felt let down by the EDI (electronic data interchange) networks that had been set up between certain major corporations. Unfortunately, the EDI investment had turned out to be less than ideal. EDI is proprietary, expensive, and slow, thus it is unable to keep up with the amount of information flowing from these high-volume manufacturers.

Most of the press RosettaNet is getting is due to the fact that it's the first attempt to implement an industry standard for XML (see Figure 2-5). But XML is only a small piece of that standard. The bulk of the RosettaNet specification centers on developing common Partner Interface Processes (PIPs) and shared dictionaries. PIPs are detailed definitions of a particular process, such as order taking. Each PIP can involve hundreds of steps and each is meticulously researched and documented. The value of RosettaNet is that its members agree on the details of particular processes, which allows each company to link common steps. Once the PIP has been constructed, manufacturers using the standard can seamlessly add new products to their partners' catalogs.

RosettaNet is based on a suite of known technologies—XML, CommerceNet's OBI specification, X12 EDI, and PKCS#7 digital signatures. Using these technologies, the RosettaNet standard defines how these technologies will behave within the ISO/OSI application and session layers (see Figure 2-6).

Other XML standards are beginning to see the light of day as well. Because of Microsoft's industry dominance, watch its BizTalk architecture. We'll discuss the new ANSI/ISA-95.00.01-2000 *Enterprise-Control Systems Integration Part 1: Models and Terminology* standard for industrial applications in Chapter 8.

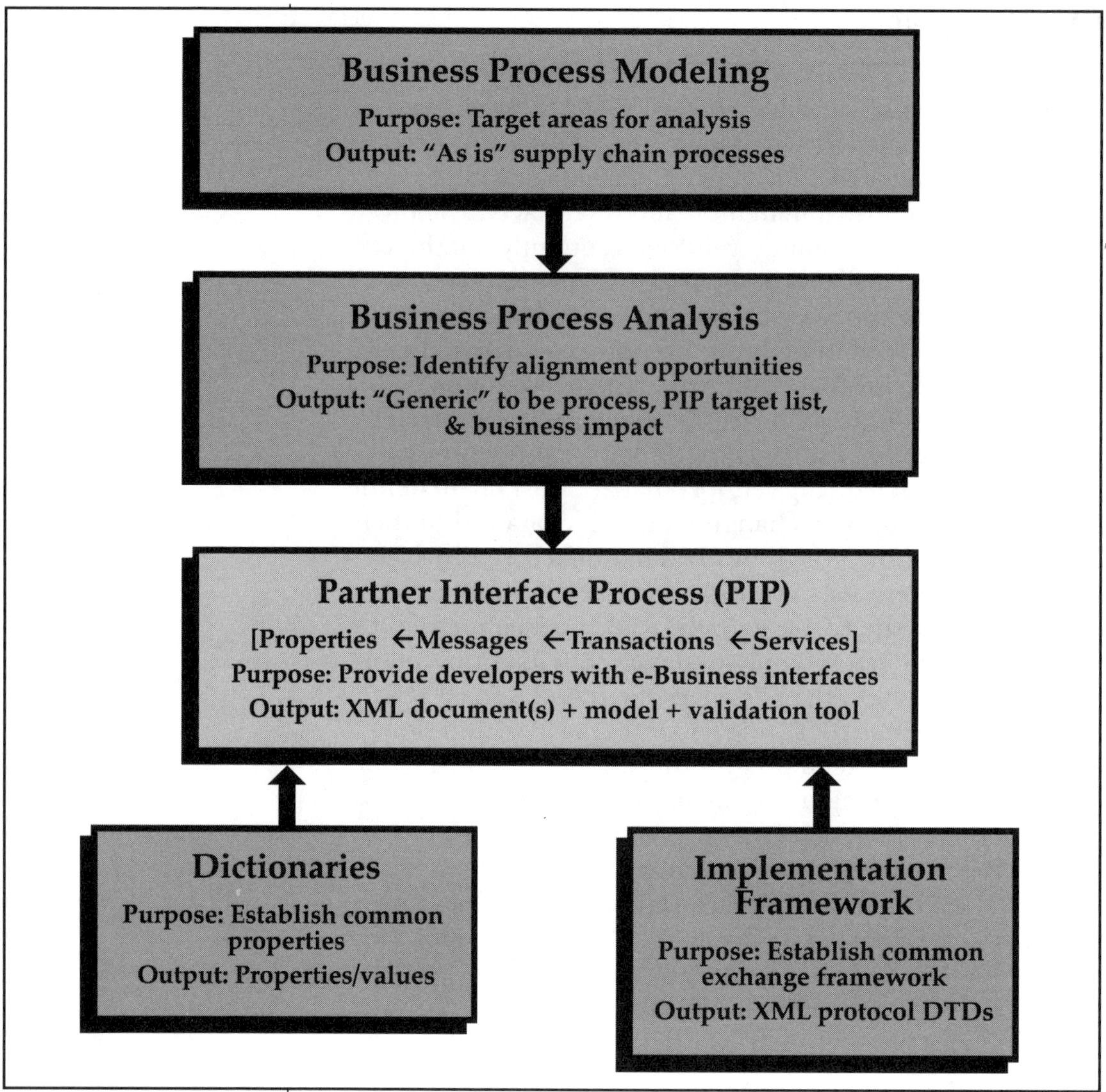

Figure 2-5. RosettaNet Development Process.

What's New in XML?

It is generally agreed that the transition from HTML to XML will be a slow one. First, it is painfully time-consuming to document the exact details of business processes, especially ones that have, for years, been taken for granted within organizations. In addition, HTML has become so prevalent across the Web that millions of sites and billions of Web pages will be affected.

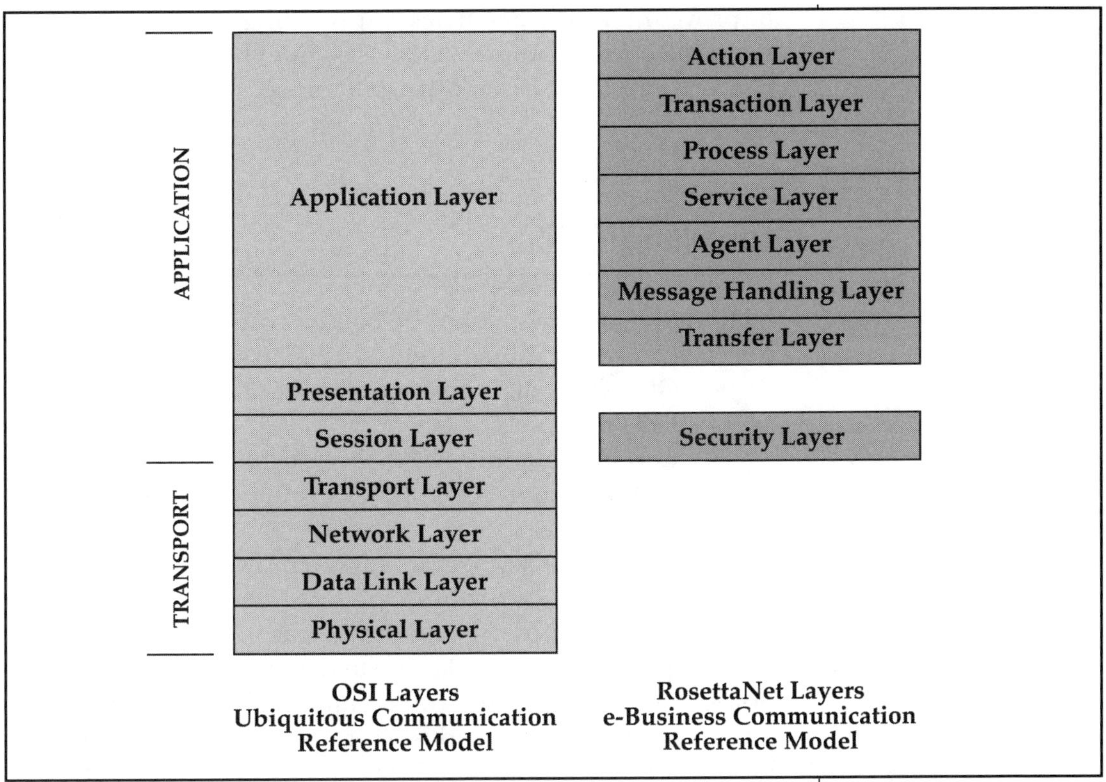

Figure 2-6. The ISO/OSI and RosettaNet Communications Reference Models.

To help with this transition, the World Wide Web Consortium (www.w3c.org) has developed XHTML, a new Web language that combines the well-formed structure and extensibility of XML with the familiarity of HTML. Like HTML, XHTML can be viewed by any number of browsers and devices. Unfortunately, XHTML is itself an ever-changing standard, something that many small and medium-sized companies can't afford to chase. How quickly these standards will settle down remains to be seen.

Intranets

The popularity of the Web and its core coding structures have had an impact that extends well beyond the Internet itself. Internal company networks are rapidly changing, and many are now being based on Internet technology. Web browsers, e-mail, and FTP have become commonplace services within companies. These "intranets," as they are called, are one of the fastest growing uses of Internet technology. Corporations are making use of intranet technology to improve internal efficiencies by providing employees with ready access to customer databases, human resources information, employee directories, and more.

EDI and Extranets

Since its acceptance in the commercial sector, the Internet has offered manufacturers a broad array of advantages:

- Global reach, global standards (TCP/IP, HTML, HTTP)
- Low cost of entry
- Flat pricing independent of volume and time of day
- Robustness
- Capable of supporting high bandwidth

Although the Internet community is producing a rapidly changing set of new standards, such as XML, its core technologies make it an attractive communications tool because it is easy to implement and is platform independent. As they have emerged, these capabilities rapidly underscored the drawbacks of EDI, the private networks that large corporations had started setting up in the 1970s to automate procurement processes with major buyers and suppliers. EDI required large capital outlays and was therefore never an alternative for small or even medium-sized corporations. When the Internet became a commonplace technology within firms of all sizes, browser-based interfaces became the online tool of choice. Instead of costly private EDI networks, companies were setting up Internet-based networks, originally called virtual private networks (VPNs).

Today, these special Internet connections between manufacturers and their partners are called "extranets." They provide manufacturers, distributors, and suppliers with fast and flexible access to exactly the data they need. Major electronics and automotive manufacturers have run dozens of extranets between themselves and their largest customers. They use these network connections to automate purchasing functions, such as allowing customers access to special pricing information. Extranets are also being used to integrate engineering teams, providing them with quick access to specifications and drawings so group interchanges between remote locations can be more efficient.

The Global Knowledge Base

The Internet architecture is based on the ideal of a global knowledge base. When the Web first emerged, its early adopters envisioned a worldwide information infrastructure. Even today, a mere handful of years after the introduction of the Web, anyone with a personal computer and modem can, within minutes, log onto the Internet and begin reading, for example, about new sensor technology from Japan.

To facilitate the rollout of such a global knowledge base infrastructure, Kalakota and Robinson suggest an organization's business logic be organized as follows:*

1. Corporate intranets: Seek to more completely and consistently link information resources scattered throughout the organization. The technology that enables this is the corporate intranet, which, using the browser as a standard interface, makes it possible to easily move from departmental solutions to a broader base of users.
2. Extranets: Focuses on supply chain partners. Fast information access, customized data, and responsiveness are key business drivers. Standardized reports and interfaces are crucial to eliminating the need for training and minimizing service requirements.
3. Commercial Internet applications: Focus on new business models. As companies gather information for their own internal activities, they should seek to syndicate and resell that information for a broader range of uses. Companies will rush to commercialize transactional data as new technologies emerge for capturing, consolidating, and reselling consumer information, business transaction records, and financial data. As a result, privacy issues, intellectual property rights, and pricing issues will gain in prominence.

* Kalakota, Ravi, and Marcia Robinson, *e-Business: Roadmap for Success.* Reading, MA: Addison-Wesley, 1999, p.296-7.

CHAPTER 3

e-Business Basics: From Customer Relationship Management to e-Manufacturing

As the Internet evolves, new businesses continually spring up, hoping to take advantage of the ubiquity and ease of use of this network of networks. Thanks to innovations like the Web, virtually anyone with a computer can traverse the Internet with a standard tool set of browser software and file transfer protocols.

Today's e-business environment enables manufacturers to automate and integrate functions as never before—from customer relationship management on the sell side, to supply chain management on the buy side, to e-manufacturing operations on the production floor.

Customer Relationship Management (CRM)

Let's start by taking a closer look at the buying side of the equation. CRM is the business function that integrates sales, marketing, and customer service. Such integration simplifies customer interaction. Within an integrated enterprise, customer content and contact information are readily available to generate additional sales and service opportunities. Integrated business processes provide consistency and simplicity to the multitude of communication channels available to firms today. And as manufacturers extend the enterprise beyond their own walls to work more closely with

The Web has expanded the reach of CRM by enabling businesses to use software analysis tools, customer interaction data, multichannel communications, and one-to-one interactions to market ever more customized products to ever more precisely defined customer segments.

partners, integration enables them to share customer information and support issues quickly and efficiently.

Traditional CRM techniques leverage call centers and direct marketing to mass-market goods and services to targeted niches. The Web has expanded the reach of this marketing function by enabling businesses to use software analysis tools, customer interaction data, multichannel communications, and one-to-one interactions to market ever more customized products to ever more precisely defined customer segments. When the Web is seamlessly integrated into the CRM program, we move into the realm of e-CRM. But there remains much controversy over just how "e"-integrated today's e-CRM products really are.

Transparency

J. D. Edwards (www.jdedwards.com) refers to e-CRM as the ability to deliver transparency to customers. This means allowing a customer to see available prices, quantities, options, alternative suppliers, and order status in real time. Most companies haven't come close to delivering such ubiquitous support, but when they do, it will provide a significant value-add (see Table 3-1, CRM in Action).

As an example, Kalakota and Robinson note that potential customers are impressed when companies call them while they're still browsing their Web site.[1] Preliminary research has shown that the probability of a sale goes up when prospects receive a response to their request for information within one to three minutes. Such is the power of closely monitored e-mail. As shown in Table 3-2, many companies are taking advantage of an integrated CRM capability.

1. Dr. Ravi Kalakota and Marcia Robinson, *e-Business 2.0: Roadmap for Success.* Addison Wesley, 2001, p. 176.

Table 3-1. CRM in Action

CRM Stage	Customer Activity	Data Source	Back-end Transactions
Information Gathering	Customer John Doe visits manufacturer's Web site for information on I/O boards	Manufacturer's Web site	Retrieve information from Web site
	John wants to know benchmarks	Testing group reports	Search for external information; results broadcast to customer
	John solicits information on new devices	Test magazines	Board articles from Test & Measurement World and Intech are e-mailed to John
Transaction Negotiation	John would like to test specific boards	Distributor database	Text to speech call is generated with data-driven voice menu to let John arrange a time for an onsite test
	John wants to review pricing	Distributor database	Distributor database queried for price information
Transaction Closure	At the distributors' Web site, John wants specifics on extended warranties	Manufacturer Web site	Manufacturer sends warranty information and customer comments to John's WAP phone
Post-transaction Service	John decides to buy the board and gives his delivery requirements	Distributor database	Information from the inventory database is recorded
	John signs all paperwork and completes the transaction	Manufacturer Web site	Existing customer information is queried to minimize paperwork
	Six months later, an upgrade to the board's drivers is released	Distributor or Manufacturer Web site	John is sent an e-mail with instructions on how to upgrade

Adapted from The Power of eCRM, a white paper by Microstrategy, Inc., 2000 (www.microstrategy.com)

Table 3-2. Fortune 100 IT Managers Are Increasingly Integrating CRM Functions into Their Web Sites

CRM Function	Percentage Integrating
Client support and services	81%
Data analysis and segmentation	66%
Promotion and ad campaign monitoring	43%
Customized messaging	41%
Other	6%

Source: Information Week magazine, ©2000 www.informationweek.com/reports

A Note on Distribution Channels

No amount of "e" anything is going to get rid of one of the workhorses of the manufacturing industries, the salesperson. In fact, the Internet will only serve to enhance the effectiveness and reach of the sales organization by providing both buyer and seller with ready access to detailed product information.

Before we go any further, let's make it clear that no amount of "e" anything is going to get rid of one of the workhorses of the manufacturing industries, the salesperson. In fact, the Internet will only serve to enhance the effectiveness and reach of the sales organization by providing both buyer and seller with ready access to detailed product information. Ultimately, sales professionals will be in a position to prequalify leads and off-load routine questions and answers that can be more efficiently handled by online information sources.

If you doubt our words, take it from Compaq Computer, now merging with Hewlett Packard, who learned the hard way not to abandon a channel that truly adds value. Early on in the Web game, Compaq became intrigued by the cost-savings potential of the virtual world. So the company began selling computers online in 1996. As the *New York Times* reported in early 2000, within a few years the company experienced a twelve percent drop in commercial PC revenues at a time when the industry was growing.[2] Company insiders admit that this drop in sales was a direct result of retaliation from resellers who objected to the manufacturer's online selling.

Never say Compaq doesn't learn from its mistakes! Compaq has since made major efforts to bring its resellers back into the sales cycle, for example, adding additional advertising and sales resources to its reseller programs. A special deal with Radio Shack allows that reseller to earn extra fees for every Compaq computer that is purchased through the special Compaq Internet kiosks in Radio Shack stores.

The lesson for the rest of the manufacturing community is that firms must carefully consider the value-add of their distribution channels as they implement their e-business

2. www.nytimes.com/library/tech/00/01/cyber/commerce/03commerce.html.

strategy. This is especially true for those companies selling complex products to sophisticated users, such as instrumentation for engineers and scientists. If resellers are more than just order-takers and are truly supporting your customers, make sure you make them an integral part of your online efforts. Start slow if you're new to this. Give resellers some space to promote their company on your corporate Web site. Forward online requests for quotes to the local reseller so they can respond directly. Even add a commission structure for e-orders. In most cases, if you work as a team your entire sales force will see the tremendous value of the Internet and will support your efforts.

e-Business Tools for CRM

In all cases, the goal is to empower one's customers to serve themselves and collaborate. This allows customers to self-qualify based on the detailed product information you make available online.

Today's CRM tools integrate traditional methods of interacting customers with automated online capabilities, including product catalogs, product configuration systems, pricing engines, proposal generators, and sales incentives and commission systems. On the customer service side, we see a variety of enhanced services being incorporated into communication channels. For instance, one of the big pushes at the moment is to seamlessly integrate real-time audio and video with Web-based data. In all cases, the goal is to empower one's customers to serve themselves and collaborate. This allows customers to self-qualify themselves based on the detailed product information you make available online. Fostering pre- and post-sales interaction between a vendor's customers also builds loyalty to the vendor because customer to customer interaction gives both existing customers and prospects hard-to-find access to innovative solutions for application problems.

"Know thy customer" is the rule of the day. As such, companies like MicroStrategy (www.microstrategy.com) have developed CRM tools that integrate personalization functions with data-mining software, transaction engines, and multimedia broadcast services (i.e., integrating Web, wireless, and voice). Most CRM solutions are geared to the high end of the market, that is, Fortune 1000 enterprises. The major players provide a standard suite of tools that encompass an increasing number of functions. At this level, we see a lot of SAP (www.sap.com), Oracle (www.oracle.com), and Siebel (www.siebel.com). Recently, perhaps because of the success of companies like Salesforce.com, Peoplesoft (www.peoplesoft.com) has led the charge to e-enable Internet-based CRM systems aimed at the small-to-medium-sized company (SMC) market.

Supply Chain Management (SCM)

Supply chain management is the business function that enables a manufacturer to manage its supply chain: the complex network of relationships between a manufacturer and its suppliers. SCM systems allow the manufacturer to coordinate a series of transactions between itself and participating trading partners, including such transactions as forecasting, purchasing, inventory status, change orders, shipment, and financing. The goal is increased efficiency through automated business processes that balance supply and demand. In a sense, SCM is another reincarnation of the just-in-time model.

Because of the global reach of the Internet and its common base of protocols, many enterprises are broadening their supply chain functionality to include Internet-enabled transactions. The Internet is an ideal vehicle for integrating what for many firms have become widely dispersed manufacturing and distribution facilities. In such an interconnected world, the Internet can draw together globally distributed islands of information into a support structure that more efficiently handles a variety of inter- and intracompany operations. Ultimately, this emphasis on efficiency enables firms to effectively fight what have become constant battles against lower margins, increased customer demands, and unpredictable sales channels.

SCM is really composed of two basic processes:

1. *Supply Chain Planning (SCP):* This includes advanced scheduling, demand forecasting, manufacturing planning, and transportation planning. All of these are necessary components for the effective coordination of manufacturing and supply efforts based on individual customer orders.

2. *Supply Chain Execution (SCE):* This process includes order planning, production, replenishment, distribution management, and logistics. All ensure that orders smoothly flow through the system, from the supplier through to the manufacturing operation and finally to the end customer.

Ultimately, SCM is about the coordination of raw materials, intermediate goods, information, and financial transactions between and among all the companies involved in producing a finished product. A variety of information and transportation links can be used to connect all these companies, essentially allowing them to function efficiently and effectively as a single organization.

What will future e-supply chains look like? Kalakota and Robinson offer insights into some of the interesting ways in which modern supply chains are evolving[3]:

1. *Integrated make-to-stock:* These are the traditional supply chains used in mass production. Production quantities and dates are forecast with no concern for individual customer needs. Orders are filled based on finished goods inventory, which leads to irregular demand flow. Examples include the supply chains of Coca-Cola, Proctor & Gamble, and General Mills. Some manufacturers, such as Hewlett-Packard, are pushing their end processes, such as adding power supplies, power cords, and manuals, to the distribution channel for fulfillment.

2. *Continuous replenishment:* To create more of a customer-demand pull system, some companies, especially in the consumer packaged-goods arena, aim to coordinate activities between the players in the supply chain more closely.

3. *Build-to-order:* In today's customer-driven landscape, the goal for many manufacturers is mass customization. It is the long-dreamed holy grail but has rarely been seen in action (save for Motorola's pioneering and hugely profitable entry into custom-built pagers back in the early 1990s). Build-to-order supply chains rely on a closer working relationship between the manufacturer and the supplier, in which information is substituted for inventory. This approach enables companies to be more flexible and therefore tighten the cost of goods in the chain.

SCM and XML

The core technology behind SCM is a combination of large, industry-standard databases, for organizing product data, and XML, eXtensible Markup Language, for exchanging data between databases. As we discussed in Chapter 2, XML is an open Internet programming standard that forces e-commerce participants to talk the same language. RosettaNet (www.rosetta.net) and UCCnet (www.uccnet.org) are leading the way in proselytizing XML by using industry collaboratives to develop guidelines and code for such vertical industries as electronic component manufacturing.

3. Kalakota and Robinson, *e-Business 2.0,* pp. 295-6.

e-Business Tools for SCM

While the big guys such as SAP, Baan, and PeopleSoft dominate SCM as much as they do CRM, a number of innovative suppliers are pushing more affordable solutions for medium-sized enterprises. i2 Technologies (www.i2.com), for example, is regarded as an innovator in coordinating materials and production throughout the supply chain.

Other vendors provide specific inventory tracking, distribution, and warehousing tools, such as the smart tags based on the RFID protocol from Philips and Motorola. Datasweep Inc. (www.datasweep.com), for example, is a developer of supply chain software that uses a Web browser-based system to enable companies to track customized products at the unit level. It then provides customers, suppliers, and the manufacturer with a real-time view into any part of the process.

ClearCross, formerly Syntra (www.syntra.com), provides companies and e-marketplaces with the ability to manage the financial aspects and logistics of their e-procurement and fulfillment operations. It sells its supply chain services on a per-transaction basis. For smaller companies, there is TheSupplyChain.com (www.thesupplychain.com), a developer of B2B SCM software that enables companies to handle procurement, fulfillment, and inventory management. The software also supports reverse auctioning, purchase orders, requisitions, invoices, and logistics.

e-Manufacturing

e-Manufacturing spans a wide range of applications, from product design and development to enterprise resource planning (ERP) and manufacturing. Thanks to the broad applicability of Internet tools and technologies, new Web-enabled solutions are widely available for use within a company for engineering and manufacturing applications.

Consider the front end of the manufacturing process. Regardless of the product being manufactured, up to eighty percent of a product's entire budget is often committed at the product concept or design stage. Additionally, up to eighty percent of a plant's conversion costs are locked in early when the product design decisions are being made. Given these major cost considerations, the availability of flexible Web-based tools to support the product design function can have a significant impact on the manufacturer's value chain. Innovations are being seen in several areas, such as real-time collaboration via Web conferencing, electronic subcontract and bid management, virtual whiteboards, even collaborative manufacturing build packages. Anything that improves a company's efficiency at the front end of the process will have a significant financial impact.

On the plant floor, the Internet's client/server architecture was quick to be applied to data monitoring applications, that is, converting remote terminal units (RTUs) into intelligent, Web-based hubs for real-time, remote viewing of production data. On the equipment side, Schneider Automation (www.modicon.com) was the first to incorporate the idea of a Web-based server into automation hardware, when it introduced a Web-based controller in 1998. Most recently, a number of industrial automation software vendors have added Web-based tools for plant-floor decision support. Wonderware (www.wonderware.com) and Intellution (www.intellution.com) are expanding from traditional SCADA (supervisory control and data acquisition) and MES (manufacturing execution system) software to decision-support tools that look at overall production and quality data in terms of order history and process conditions, alarm analysis, even work-in-process tracking. This infringes on what has traditionally been the exclusive domain of the plant-floor information vendors, developers whose software resides in that ethereal layer between ERP and MES. Ultimately, the distinction between plant floor, MES, and ERP has become hopelessly blurred.

Tying the plant together, enterprise resource planning (ERP) and enterprise asset management (EAM) tools integrate plant-level information with the financial information needs of the business. Using Internet technology to more effectively integrate plant information is a central unifying theme for the twenty-first-century manufacturing enterprise. Such software encompasses accounting and finance, production planning and material management, and human resources. Moreover, it can be integrated with internal CRM and SCM systems and provides comprehensive, enterprise-wide reporting.

"Manufacturing Execution Systems deliver information enabling the optimization of production activities from order launch to finished goods. Using current and accurate data, MES guides, initiates, responds to, and reports on plant activities as they occur. The resulting rapid response to changing conditions, coupled with a focus on reducing non-value-added activities, drives effective plant operations and processes. MES improves the return on operational assets as well as on-time delivery, inventory turns, gross margin, and cash flow performance. MES provides mission-critical information about production activities across the enterprise and supply chain via bi-directional communications."
© 2000 MESA

e-Business Tools for e-Manufacturing

Major developers in the industrial automation market are adding Web-based functionality to an increasing variety of plant-floor equipment and automation software. At the moment, you'll find that traditional players, such as Fisher Rosemount, ABB, Invensys, PeopleSoft, and Rockwell Automation, are spending as much time adding "e" capabilities to their equipment as they are trying to figure out which e-marketplace will best serve their CRM and SCM needs.

As in most markets, innovations often surface first through smaller developers. As early as the ISA/2000 conference in New Orleans, we saw bleeding-edge demos from software companies like LABTECH (www.labtech.com) who showed plant-floor data being transmitted in real time over cellular phones and palm-sized computers. Browser-based everything became the norm a mere three years after the emergence of the Web: from Web-enabled SPC from DataNet and browser-based CAD viewers from Bentley Systems to Web-based data integration software from Lighthammer.

Online Exchanges

Online exchanges began to spring up soon after the emergence of the Web. In the manufacturing sector, the first stage of development was so-called portal sites, or online knowledge bases of information about where to find products and technology specific to vertical industries. These portals then developed into a new distribution channel for manufacturers and their products. These exchanges sought to take advantage of the low cost of electronic transactions by serving as intermediaries to manufacturers who couldn't afford to set up their own e-commerce sites or who were looking for broader exposure to potential customers. Such online exchanges, which often included auction capabilities, became known as electronic marketplaces or "e-markets."

Two basic types of e-markets are currently battling for market dominance: third-party exchanges, in which a neutral company brings buyers and sellers together online, and industry consortia, in which the major players in a particular market sponsor, develop, and support an online exchange for buyers and sellers. At this point, it is unclear if any type of e-market will survive as a profitable business model for the manufacturing sector since several major players have already admitted that transaction fees alone do not a profitable business make.

Defining e-Markets

In 2000, *Forbes* magazine took a decent stab at making sense of the rapidly changing e-marketplace space.[4] Their

4. "The B2B Universe Explained," *Forbes*, July 17, 2000, p.153-55.

editors broke the business down into the seven types shown in Table 3-3.

According to Arthur Sculley and William Woods,[5] four trading methods are commonly used in today's e-marketplaces:

- Fixed pricing, such as catalog resellers use
- One-to-one negotiation
- Auctions (both seller-driven and buyer-driven, or reverse, auctions)
- Automatic systems (two-way auctions)

Table 3-3. A Diversity of e-Market Models Are Sorting Themselves Out

Type of Exchange	Characteristics
Buyer Exchange	Dominated by the large companies in a market. Suppliers bid to fill contracts at the lowest price in reverse auctions, e.g., Covisint, FreeMarkets, Petrocosm.
Supplier Exchange	Industry vendors come together to sell their goods as a group, e.g., GE Polymerland, ChipCenter, MetalSite.
MRO-Catalog Hub	Suppliers sell nonproduct-related items used for maintenance, repair, and operations, such as pencils, furniture, e.g., Grainger.com.
Neutral Exchange	Independent market for goods and services, e.g., PaperExchange.com. We would have mentioned Chemdex.com here, but they hung up their browsers in December 2000 after two years of what they had consistently claimed was profitable business.
Technology Enablers	Companies that provide the software apps and B2B marketplace expertise, e.g., Ariba, CommerceOne.
Financial Enablers	Groups that provide finance and other resources for e-commerce, e.g., Internet Capital Group, Accel Partners
Business Process Outsourcers	Hubs and/or application service providers (ASPs) that manage noncore functions for companies and exchanges, such as human resources, payroll, and facilities management, e.g., Cadence Network and Employease.

MyPlant.com

In the instrumentation and control equipment industry, we are seeing the development of both third-party and vendor-sponsored exchanges. The most obvious example is MyPlant.com, an industrial automation portal developed by Honeywell Industrial Automation (iac.honeywell.com). Looking to garner wider industry support from a variety of equipment vendors, and concerned that this new sales channel might create a feeling of conflict for existing Honeywell distributors and value-added resellers, the site is about to become an independent organization.

5. Arthur Sculley and William Woods, *B2B Exchanges: The Killer Application in the Business-to-Business Internet Revolution.* Addison Wesley, 2000.

Company officials describe MyPlant.com as an "e-hub," another term for e-marketplace. But they claim to be an e-market with a difference. For MyPlant.com, the actual selling of equipment is peripheral to the primary goal: delivering knowledge-based solutions that integrate software, hardware, services, and information for engineers in process (and soon also discrete) manufacturing industries.

MyPlant.com is attempting to differentiate itself by offering such capabilities as application service provider (ASP) services, which include functions like delivering remote diagnostics or process modeling software tools in a "pay-as-you-go, pay-for-the-time" manner. Additionally, the company has added an artificial intelligence engine that answers users' questions about process manufacturing. This part of the service is based on the AskJeeves.com technology, which searches for and provides answers to problems, learning as it goes.

At this point, it is unclear if MyPlant.com will be able to survive without the direct intervention of its benefactors' deep pockets (in addition to Honeywell, owners include Microsoft and VerticalNet). Income at MyPlant.com is based on advertising revenue and commissions on sales. Now that it is widely acknowledged that Internet businesses entail high overhead, these sources may not be enough. Only time will tell. In the meantime, however, MyPlant.com has struck a chord with industry pundits with its emphasis on "solutions."

Other instrumentation vendors that are attempting to make it with online markets include ABB Automation, which has teamed up with PaperLoop.com to sell pulp and paper products; Rockwell Automation, which launched SourceAlliance.com as an independent company; and Siemens' Vertacross.com. Third-party players in this market include TestMart.com and the VerticalNet portals (which encompass several dozen niche sites, including ElectronicsWeb.com, PlantAutomation.com, and TestandMeasurement.com).

Beyond Transactions

So how did the whole e-marketplace hysteria begin? As in many other cases, such as the push for ISO and, dare we mention it, reengineering, the concept was driven forward by the promise of vastly improved productivity. In the case of e-markets, the incentive was the hope of increased riches from streamlining production and supply. A constant supply of breathless statements from industry analysts only served to enhance the already frenetic pace at which these new channels were evolving.

For instance, Deloitte Research proclaimed in October 2000 that "Manufacturers can be up to seventy percent more profitable by connecting with all trading partners through the Internet. Those companies that are able to integrate their supply chain and their customer strategies will achieve breakthrough performance." But here's the catch—even though strong benefits can be gained from integrating the supply chain with customer relationship management (CRM) tools, research shows that only ten percent of businesses have the right kind of internal and external networks to provide any kind of significant integration.

Similarly, in the autumn of 2000, Jupiter Research projected that "U.S. B2B e-commerce revenues are set to take a huge leap, growing from $336 billion in 1999 to $6.3 trillion in 2005. Online supply chains will dominate the landscape, so companies are advised to *adopt new models for buying and selling* as B2B commerce swells from three percent to forty-two percent of domestic B2B trade."

It's not surprising then that a number of "e-enlightened" entrepreneurs jumped at the chance to stretch traditional concepts of distribution by taking advantage of the technological ease with which communities can be established online. In the case of e-markets, we're talking about communities of buyers and sellers who are looking to ease the time and expense of managing their supply chains.

e-Markets: What Went Wrong

But something's not quite right with e-markets. To begin with, most focus on price as the primary criterion for attracting buyers and sellers. For buyers, these online exchanges offer a quick way to get product and price information from a group of vendors. While this approach is convenient, it can mislead buyers into searching less vigorously for the products they need since not all vendors are going to embrace the e-market model. How easy is it to find a 24-bit data acquisition board that supports a 1MHz per channel sampling rate?

For the vendor, the e-market's emphasis on SKUs (stock keeping units) and prices is a danger sign. Without a way to sell value-add, vendors risk a quicker path to commoditization. Then there's the issue of how the e-marketplace financially supports its business. Most such ventures focus on facilitating purchase transactions. This means they list goods and services for sale from a variety of vendors, allow buyers and sellers to negotiate on price, and execute the placement

For the vendor, the e-market's emphasis on SKUs (stock keeping units) and prices is a danger sign. Without a way to sell value-add, vendors risk a quicker path to commoditization.

of the order. But most e-markets significantly underestimate the time and cost of integrating dissimilar systems. Exchanges assume that they can make it up in volume by spreading the trading technology across multiple industries or niches. Unfortunately, as of yet, you can't just plug these things in and expect them to work.

A complicating factor is that the bulk of B2B transactions are, and for some years will continue to be, mixed mode. This means that although credit cards can be processed online, many transactions must be handled off line by staff who are in direct contact with buyers. Such transactions include accounting, legal advice, custom product specification and design, even negotiated contract terms. Rich-media technologies, such as voice and video over IP, will benefit this effort, but these technologies are still in their nascent stages.

This leaves e-markets relying solely on transaction fees for their revenue base. With the inevitable and constant push on margins, however, such fees alone will never support a technology-intensive business like an e-market. In late 2000, several independent markets realized this and started to add e-services similar to those offered by MyPlant.com. Such services included software licensing, e-tools integration, and logistics support (e.g., carrier connectivity and freight forwarding).

Established exchanges redefined themselves, struggling to find a profitable business model. SciQuest.com, a leading e-market for laboratory supplies, redefined itself as an integrator of private e-markets. They hope this will help them put their e-commerce experience to profitable use.

For both buyers and sellers in the manufacturing markets, certain needs are obvious but not yet met:

- Collaborating on design for more complex products, including the ability to exchange and edit CAD files
- Personalized content to help customers quickly and easily learn about technologies and products crucial to their business
- Contract and resolution management, a complex process in most B2B transactions

What We've Learned from Financial Services

Ultimately, exchanges will not survive in their current form because they are focused on price and, as such, do not address the real needs of the manufacturing markets. Many companies have expended much time and effort moving beyond a price-only mentality, especially in this day and age of six-sigma quality. Today, we find manufacturers developing tighter relationships with suppliers and emphasizing joint product design, integration of complex processes, and long-term service contracts. Instead, e-markets focus on arm's length, price-driven transactions.

In their *Harvard Business Review* article, "Beyond the Exchange: The Future of B2B," Richard Wise and David Morrison point out that several factors are reshaping B2B.[6] Among these are more efficient pricing, more effective matching of buyers and sellers, and, most significantly, the shift of value from the product itself to information about the product. Wise and Morrison propose that the emerging B2B landscape will be comprised of a series of industry players that are geared to handle the various pieces of a "solution sell." Solution selling is a practice quite familiar to the manufacturing markets, but Wise and Morrison propose that we will see the financial services industry, which has already paved the way, exert heavier influence on accelerating this trend.

As Figure 3-1 shows, in the new B2B e-commerce picture "mega-exchanges" will act as central hubs for executing most of the transactions and commun-ications between buyers and sellers. "Specialist originators" will emerge that offer strong consultative sales skills, deep product understanding, and strong customer relationships (sound like a distribution channel you may already know?). They will use the Internet to standardize and automate the buyer's decision-making process and then send the transactions to the exchanges for execution. These exchanges will be a bare-bones version of what we see in the market today. In the near future, exchanges will run on slim or no profits, primarily serving as a collective that benefits the entire industry.

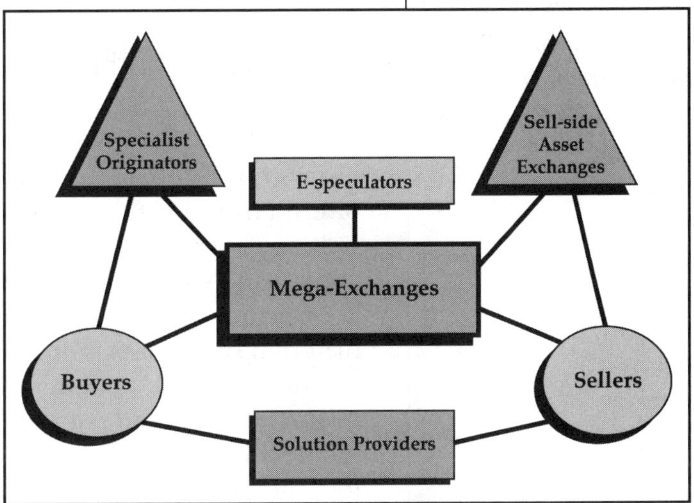

Figure 3-1. The future of e-business in manufacturing revolves around a core hub that handles purchase transactions. Solution providers continue to add value with their industry expertise. (Source: Harvard Business Review, November/December 2000, p. 93)

"Solution providers" will also play a role in the new B2B e-commerce by embedding the product sale within a suite of unique and valuable services. "e-Speculators" will also participate by buying and selling large quantities of the products traded in their markets. Enron, for example, originally a gas pipeline operator, is now a major provider of other energy

6. Richard Wise and David Morrison, "Beyond the Exchange: The Future of B2B," *Harvard Business Review*, November/December 2000, pp. 87-96.

products. The company operates an online exchange that makes money from buying and selling a variety of energy products for its own accounts, including natural gas, electric power, pulp, and pollution credits. As an e-speculator, Enron now creates and sells derivatives of these products, including options, futures, and swaps, which allow other market participants to hedge their price risks.

"For many companies, traditional skills in such areas as product development, manufacturing, and marketing may become less important, while the ability to understand and capitalize on market dynamics may become considerably more important." Wise and Morrison, Harvard Business Review, Nov-Dec 2000.

According to Wise and Morrison, "For many companies, traditional skills in such areas as product development, manufacturing, and marketing may become less important, while the ability to understand and capitalize on market dynamics may become considerably more important." [7] In the case of Enron, for instance, as the company expanded its skill set it brought in new people, shifting from mostly engineers to a mix that included financial traders, economists, and risk managers.

e-Business Models

So where does this leave you, the manufacturing firm or distributor? Well, to best take advantage of the possibilities offered by the Internet, your organization must start with a viable strategy. The experts each have their own theory about which approach will best anticipate future needs.

Gartner Group (www.gartner.com), an industry analysis firm that specializes in information technology, is espousing a concept called "collaborative commerce." Made possible by the Internet and component models of product development, collaborative commerce ("c-commerce") brings together a manufacturing firm, its business partners, and its customers in what has long been referred to as a "virtual enterprise." Today's Internet technologies provide a common platform for inter-enterprise collaboration, from product design to production to sales. By 2005, Gartner expects c-commerce to be the dominant model for e-business.

According to Bruce Bond, senior analyst at Gartner, "c-commerce is aimed at creating a collaborative framework that allows companies to engage with cyber-market partners for the purpose of creating and connecting agile business processes. e-Commerce is designed to construct a virtual link between a pre-defined community of trading partners for the purpose of buying and selling goods and services. Content is generally confined to Web catalogs of finished goods and collaboration mechanisms center on exchanging mes-

7. Wise and Morrison, "Beyond the Exchange," p. 95.

sages or purchasing transactions. c-Commerce provides a much deeper and richer form of B2B interaction designed to allow trading partners to serve as virtual collaborators across a wide array of business processes."[8] Table 3-4 outlines the phases of this transition to a collaborative environment.

Table 3-4. The Transition from e-Commerce to c-Commerce

	e-Commerce	c-Commerce
Content	Product catalogs	Rich information assets
Community	Predefined trading partners	Cyber market
Commerce	Buy, sell, auction	Collaborative business processes
Collaboration	Exchange messages	Exchange intellectual capital

Source: Gartner Group presentation, November 2000

Gartner makes a strong recommendation: Enterprises trying to choose between competing public marketplaces should favor those with strong collaborative capabilities and c-commerce visions.

Developing an e-Business Strategy

Companies need to adopt a comprehensive approach to fully embrace the interactive and ubiquitous capabilities of the Internet. Simply applying old habits to new technologies will not do your e-business initiative justice and will more than likely leave you wondering where the profit potential comes into play.

Understanding how to integrate your business needs with the potential of the Internet starts with a thorough analysis of your customer base, your internal capabilities, and your goals. For most manufacturers, the goal is to maximize customer value and, along the way, turn a profit. Kalakota and Robinson have developed a comprehensive three-phase checklist for companies developing an e-business strategy.[9] How you answer these questions can make the difference between profit and loss in your e-business endeavors. If you have trouble answering them, call in a specialist. A number of competent firms have the requisite knowledge to help you develop your e-business strategy.

8. Gartner Group presentation, November 2000.
9. Kalakota and Robinson, *e-Business* 2.0, pp. 395-414.

Phase 1: Knowledge Building

- Understand the customer: Who are my customers? How are my customers' priorities shifting? Who should be my target customer? How will the e-business help reach my target customer segments?
- Customer value and relationship trends: How can I add value for the customer? How can I become the customers' first choice? How does my product reach the customer?
- Technology trends: Do we understand the environment and industry trends? Do we understand the technology trends?
- Supply chain trends: What are the priorities in the supply chain?
- Competition: Who are my real competitors? What is my toughest competitor's business model? What are they doing really well?

Phase 2: Capability Evaluation

- What internal capabilities do we have today?
- What capabilities and resources do we need to execute quickly?

Phase 3: e-Business Design

- What's my e-business design—category killer, channel reconfiguration, transaction intermediary, infomediary, self-service innovator, supply chain innovator, or channel mastery?
- Refine the design. Look at customer selection, customer experience, customer capture, scope of design, ease of doing business, and organizational systems.
- Clarify the differentiation levers. What is my market positioning (how do I want to compete for customers in the market) and differentiation strategy? Base your choice of product features, marketing channels, service and support, and brand or image positioning on critical selection criteria or on price.

Technology Convergence

Sparked by the Internet, a remarkable convergence of technologies has hit the manufacturing enterprise. Embedded computing is combining with the Internet to produce truly intelligent factory-floor devices. Before long, these devices will be self-maintainable, that is, able to diagnose an upcoming repair or maintenance operation without human intervention.

Wireless technology and the Internet are combining to support any-to-any computing. This means any person with proper authorization can have ready access to virtually any form of corporate information available, from high-level profit and loss data to sensor readings from plant-floor equipment.

Ultimately, the Internet empowers virtually everything it touches, from people to factory- floor equipment. Software agents are making remarkable progress after decades of fits and starts. In the near future, automatic, computer-generated interactions will free users from virtually all mundane tasks and bind customers to vendors through powerful e-business architectures.

Peer-to-Peer Networking: The Impact of Music on Manufacturing

Covisint, e-Steel, CheMatch.com...the prevailing model of B2B e-commerce centers either on solo corporate sites or e-marketplaces. For those who can't afford a big solo effort or who are looking for broader exposure, large exchanges bring buyers and sellers together through either a third-party intermediary or an industry consortium. These exchanges look to facilitate and increase the efficiency of trading partners by providing them with a common platform of software tools and e-services. It's a compelling concept, but as we're already seeing, it's one that may be past its prime before its prime was even really reached. So what are we left with? Back to the solo corporate site? Maybe not.

What started out as a little piece of the music industry promises to change the face of e-transactions. Peer-to-peer (P2P) networking, which has been available on the plant floor for decades, has now hit the Internet. Participants in a peer-to-peer network, either individuals or corporations, exchange information directly with each other, much like a data acquisition system interacts with its remote terminal. There is no need for a central exchange, as is typically the case in the client/server world of the Internet.

Some P2P networks resemble the one created by Napster, the music distribution Web site that nearly brought the music industry to its knees in the late 1990s. In this variation, information about the location of files is stored in a central directory, but the files themselves are located on individual users' computers. In other scenarios, there's no central directory at all; each computer on the network simultaneously serves as client and server.

P2P in e-Commerce

P2P networks provide some interesting advantages to the world of e-commerce. As Andrew McAfee noted in *Harvard Business Review*, "First, and most obvious, they allow companies to avoid the fees charged by exchanges. Second, they reduce the complexity and expense of networking. It's much easier for a company to integrate its internal information systems with a single P2P program, or 'peering portal,' than with a bunch of different exchanges. Finally, while membership in exchanges is limited, often determined along industry lines, P2P networks have no bounds. A company can search as widely as it wants for new partners and products or for lower prices or better terms." *

The trick here, though, is standardization. Companies still have to agree on a standard method for describing product specifications and business processes. This enables them to exchange information without human intervention on each and every transaction, such as searching inventory and tracking shipments. For now, we have the infamous browser and HTML. But that leaves much of the processing to humans who must manually force their way through a vendor's ordering system. New Internet technologies like XML and UDDI (Universal Description, Discovery, and Integration**) show promise, but they require that partners agree on how any specification—product, production plan, purchase order, contract—will be described electronically. P2P networking will play a role in the future of e-commerce, but not until it addresses the biggest issue holding us back—standardizing information transfer among buyers and sellers.

* Andrew McAfee, "The Napsterization of B2B," *Harvard Business Review*, November/December 2000, pp. 18-19.

** UDDI (Universal Description, Discovery and Integration) is a technology pioneered by Microsoft and IBM to establish a global directory where companies can post such specifications in a standardized format that allows easy integration into XML.

CHAPTER 4

Security and e-Commerce

Since its inception, the biggest hindrance to widespread acceptance of the World Wide Web and e-commerce has been a fear of less-than-perfect security. This issue surfaces in retail e-commerce in consumers' fear that someone will steal their credit card number and make purchases without their knowledge. Though they are in fact rare, several highly publicized "break-ins" have occurred in which hackers have stolen credit card numbers from large online vendors and credit card companies. In industry, the most common online security fear is that someone will steal vital company secrets and sell them to competitors or release them to the press.

Other security issues impact both retail and the industrial establishments, such as the unnoticed virus that slips onto a computer and destroys large volumes of irreplaceable data. In other instances, when hackers are unable to break through security barriers, they may resort to "denial-of-service" attacks that flood the target server with data until it crashes under the heavy burden. Another type of security breach is site "takeovers;" these have defaced both government and business Web sites. In one such example, a group of hackers from China targeted U.S. servers after a minor confrontation between reconnaissance aircraft. Regardless of the form online security threats take, the danger is quite real and must be addressed, even though the hysteria over such occurrences is often overblown.

The world of Internet security moves and changes amazingly fast. For that reason, it is difficult to go into as much detail in this chapter as we would like. So in addition to our brief discussion of Internet security here, we've included a list of periodicals and Web sites that deal regularly with Internet security issues in the Glossaries.

Enterprise-wide Security

The extended enterprise has many interlocked parts. Each faces its own security issues and thus poses a potential threat to security through all the company's interconnected servers. For example, let's look at human resources (HR). If HR is a closed system and not part of the company's internal network, there are few security issues. However, as soon as you open the HR system to the company intranet, potential threats arise. Can a factory-floor worker use a stolen password to gain remote access to a manager's salary records? Can a corporate partner who is visiting the factory for the day access employees' home addresses by tapping into the internal network when he retrieves his external e-mail?

Such security issues go far beyond HR. Also vulnerable is the extranet through which distributors, salespeople, and customers access account information, inventory status, and shipping information is also vulnerable. Even a company's e-mail system is susceptible. For instance, if you are working on a Windows NT or Windows 2000 system, you face potential security holes both present in the OS itself and introduced through externally introduced, script-based "viruses." But you are no safer if you run in the UNIX or Linux environment, though it is true that most e-mail-borne viruses are aimed at Windows-based systems. (Why aim to hit one million Macintoshes if you can hit forty million Windows machines at once?)

"In the physical world, security is implemented to manage risks, not to try to eliminate them. Grocery stores accept that some shoplifting will occur, but try to compensate for it by using security devices, employees, and insurance. Computer security must adopt the same stance." Bruce Schneier, Counterpane Internet Security, Inc.

The solution to this multiplicity of threats is a balanced approach to managing risk. While nothing can compete with the perfect security of a computer that's unplugged and buried underground, there is a realistic compromise. Bruce Schneier, CTO of Counterpane Internet Security, states, "In the physical world, security is implemented to manage risks, not to try to eliminate them. Grocery stores accept that some shoplifting will occur but try to compensate for it by using security devices, employees, and insurance. Computer security must adopt the same stance."[1]

1. Sam Costello, "Internet Too Complex to Secure, Says Exec," IDG News Service, July 13, 2001 (www.nwfusion.com/news/2001/0713intsecure.html).

Analyzing Risks

The biggest risk is human error. In the daily rush of business, in the press to get work done, and move on to the next task, sometimes even the most experienced IT professional will sometimes make a mistake, even if it's only continuing to keep the username "Admin" and password "root" after system startup. Human error is unavoidable, but it can be planned for and its effects can be mitigated. David Thompson, an independent security analyst formerly with PricewaterhouseCoopers, recently wrote that "Security management today consists of mostly putting out fires. You do what you can, hope you haven't missed something, and pray that today isn't the day an overlooked BIND vulnerability is used to hack into your company."[2] Thompson provides a useful checklist of best practices for mitigating security problems:

- Get everyone on board
- Develop a security infrastructure
- Establish a training and awareness program
- Develop threat sources
- Don't try to protect everything
- Outsource carefully
- Develop a response plan
- Perform regular audits

"Security management today consists of mostly putting out fires. You do what you can, hope you haven't missed something, and pray that today isn't the day an overlooked BIND vulnerability is used to hack into your company." David Thompson, PricewaterhouseCoopers

Getting everyone on board may sound good, but it isn't easy. It's also not without costs. Many companies face an even more interesting problem in that their systems have evolved from many sources and bring with them all sorts of legacy hardware and software. Determining "permissions," approval levels, and the admin hierarchy are difficult tasks when you are faced with the competing needs of many disparate departments and divisions. Some resources may even be located in different countries. Many IT managers complain that merely getting people to change their passwords regularly hard enough. Forcing them to do so creates resentment against the IT department, but not pursuing the issue creates an enormous hole in security.

2. David Thompson, "The Social Engineering of Security," *eWeek*, June 11, 2001.

Security Training

Making everyone aware of the threat is the purpose of a training and awareness program. Every corporation should regard the need to train employees about security as equal in importance to developing the security tools. If people aren't trained to recognize the need for security, they will not follow guidelines and implement procedures because they see as just another task to do.

The most common threat comes from disgruntled employees or former employees who discover that they still have access and permissions. By employing vigilance and good password and permissions schemes, it is fairly easy to defend against the actions of former employees. It is less easy to defend against the actions of disgruntled employees, since you must presume that they have a right to the passwords and permissions they have.

Thompson argues that developing a list of threat sources is critical to the success of any security program. The most common threat comes from disgruntled or former employees who discover that they still have access and permissions. By employing vigilance and good password and permissions schemes, it is fairly easy to defend against the vengeful acts of former employees. It is less easy to defend against the actions of disgruntled employees since you must presume that they have the right to possess the company's passwords and permissions. The best defense relies on comprehensive employee training and good employee relations. If your employees are loyal and happy, the number of incidents caused by "inside activity" will diminish to a negligible level.

Unless you are a financial institution or a large retail establishment, you probably won't be hacked for credit cards. However, you might be hacked by a person or group with an agenda that includes anti-industry sentiment. Pharmaceutical manufacturers, energy companies, military contractors, and genetic engineering firms are vulnerable to politically motivated agendas.

Once you have determined which are the weakest points of your system and who is likely to present a threat, you can defend against those specific threats rather than vainly try to defend against everything.

e-Mail and Virus Threats

The most formidable day-to-day threat to the extended enterprise is the continuous bombardment of virus-laden e-mails on the organization's servers. These e-mails have become quite sophisticated. Exploiting the security holes in e-mail clients as well as in Microsoft Outlook, MSMail, and Microsoft Exchange (the most commonly used e-mail clients in the business world), these viruses are disguised as non-threatening files by being attached to "reasonable-sounding" messages.

For example, Windows 98/NT/2000 defaults to "file extensions not shown." As a result, virus writers found that it was possible to disguise a virus by giving it an inane name such as "OfficePhoto.jpg" when the actual file name is "OfficePhoto.jpg.pif." ". PIF" is an executable file extension, similar to ".doc" and ".exe", but you don't see the last three characters. As a result, you think you are opening a photograph when in reality it's an executable file that filters its way through some or all of your hard drive, looking for ways to propagate itself to other computers. Most insidiously, more recent virus e-mails quote your own post back to you, come from an address you actually sent the post to, and have the comment, "look to the attachment." Unwary or busy people click on these virus-laden attachments with devastating results, especially within a corporate network. If the company is using a networked version of Outlook 2000 or Outlook XP, the infection of a single network workstation can cause the infection of the entire Outlook database on every workstation in the network, usually before the systems administrator can stop it.

Some companies have become so paranoid that they have set corporate-wide filters that prevent any attachments from passing through their firewalls other than simple text (DOS text, or .txt files) or rich text (Windows and Macintosh text, .rtf files). One company, a large Canadian oil producer, quarantines all attachments other than text for forty-eight hours. A systems administrator then manually checks each attachment for viruses before delivering the e-mail. It's easy to imagine the drain such an approach has on IT's productivity.

Unfortunately, there is no way to completely defend against viruses transmitted in this manner given corporations' current dependence on Windows software. The best approach is both to have an active virus checker at each workstation, such as McAfee Virus Scan or Norton Anti-Virus, and to run an active training program that ensures employees don't defeat the system just for the sake of personal convenience.

Viruses, Trojan Horses, Cookies, and Worms

Ever since November 2, 1988, when the Great Internet Worm was first released, Internet users have been wary of "hacking." Originally a name for good code writers, the term *hacking* has become, along with *cracking*, come to mean the practice of applying malicious programs to someone else's computer or network. The most common types of malicious programs are viruses, Trojan horses, cookies, and worms.

Viruses are self-replicating bits of information that derive their name from their biological counterparts. They can be injected into a computer's operating memory and used to perform actions without the knowledge of the computer owner or user knowing. Viruses can cause the individual computer to malfunction by erasing the FAT table on the hard disk or deleting important operating system files. Viruses that use subroutines to self-replicate and infect other computers or networks are referred to as "worms." The name comes from the type of program described in the 1975 science fiction novel *The Shockwave Rider* by John Brunner as well as even earlier works by Walt and Leigh Richmond. Whether Brunner and the Richmonds were describing existing work in programming or programmers tried to duplicate their concepts in the real world is a matter of debate.

Trojan horses are bits of code maliciously inserted into otherwise normal programs that can activate themselves or be activated remotely. The PKZIP.EXE Trojan horse is the most obvious example: a Trojan horse was inserted into the most popular download file on the Internet at the time, PKZIP. Users who downloaded the infected file and installed this version of PKZIP found their computers trashed and their files destroyed.

Normally, "cookies" are beneficial bits of Internet code that allows Web site visitors to register their contact information and login data with a Web site. The Web server keeps this information in a database and retrieves it whenever the person returns to the site. This allows site visitors to access all site information without the inconvenience of repeatedly entering the same data.

Normally, "cookies" are beneficial bits of Internet code that make it possible for Web site visitors to register their contact information and login data with a Web site. The Web server keeps this information in a database and retrieves it whenever the person returns to the site. This allows site visitors to access sections of the site without having to repeatedly enter the same data. Amazon.com is the most notable example of a site that relies on cookies. The online bookstore uses them to match your computer with a database record that keeps track of your name, reading preferences, purchase history, and credit card information. Cookies have become ubiquitous since 1996 because they make Web commerce much simpler and straightforward. By storing a few bits on your computer, sites can customize and personalize themselves for you, making your Web interactions that much more useful.

At its worst, the cookie usually only annoys people by cluttering up their hard drive with special tags. Unfortunately, even this fairly innocuous tool increases the potential for security breaches of Web servers and users' computers. For example, malicious cookies can be secretly inserted into a user's cookie file when he or she visits a Web site. As a result, some companies require their employees either to turn cookies off or to set their browser's permissions to accept each individual cookie as it is downloaded. That latter approach enables you to know when a cookie is being loaded onto your hard drive; you can then refuse it if it's from a site you do not want to register at. Vanderbilt University has produced an excellent report outlining the risks and benefits of cookies in commercial enterprises.[3]

According to worm expert Bob Page, "A worm is a program that can run by itself and can propagate a fully working version of itself to other machines."[4] The Great Internet Worm of 1988 brought sharply into focus the fact that UNIX (and other "–nix"-like operating systems such as Xenix, SCO-UNIX, Linux, and Mac OSX) was designed for ease of use, not for unbreakable security. The same is true even for Microsoft's Windows operating systems. Worms can be used to deny service, perform functions, and replicate themselves. Many so-called MS Word macro viruses are in fact worms.

Denial-of-Service Attacks

In recent months, we have seen a new kind of attack on government and corporate Web sites based on the "denial-of-service (DOS)" or "distributed denial-of-service (DDOS)" method. The most disconcerting aspect of DOS attacks is that relatively unskilled people can employ them by using third-party "hacker" tools readily available online. Such attacks are difficult to defend against because they do not attempt to penetrate your system. Instead, they continuously "ping" your server.

A ping is a software command that sends a short packet of information over the Internet to check the operating condition of your server. Software utilities, such as www.247sitewatch.com, regularly check the "uptime status" of your Web site by applying the "ping" command. If you repeatedly ping a Web site's URL or a DNS address over a short period of time, you can prevent real traffic from reaching that Internet node. This effectively "denies service" to that URL or DNS address.

3. ecommerce.vanderbilt.edu/cb3/mgt565a/group5/paper.group5.paper2.htm.
4. Bob Page, "A Report on the Internet Worm," November 7, 1988, University of Lowell, found at www.goldinc.com/html/maloy/SECURITY/morris_worm.html.

Distributed DOS attacks use multiple computers to accomplish the same result, making it even harder to trace the origin of the attack. Any computer that resides on the Internet can be subverted and taken over by a "bot" designed to perform distributed DOS attacks. A bot is a "robotic" software program that uses a computer's operating system, memory, and drives to repetitively complete an action (hence the name "bot"). Pinging a remote server is a very simple bot program, and many such bots are available at hacker sites across the Internet.

Companies like Microsoft, several U.S. government agencies, the Israeli government, and computer security consultants like Steve Gibson of Gibson Research Corporation have all been subjected to this technique. It has been used by politically inspired groups, disgruntled employees, even children. One attacker currently being investigated by the FBI is a thirteen-year-old! At the time of this writing, Steve Gibson's detailed history and analysis of the DDOS attack that crippled his company's Web presence in June 2001 could be found at www.grc.com. Especially disconcerting is Gibson's assertion that Windows XP (scheduled for release in late 2001) contains tools that will make it easier to arrange DDOS attacks and harder to find their perpetrators.

The only way to keep your site online in the face of a DOS attack is to work with your ISP to filter the DOS traffic before it reaches your server. This can be difficult to do, but because the chance of DOS attacks is quite real such precautions should be a critical part of your security system planning.

Firewalls and Routers

The best system architecture for protecting an enterprise against external attack is a series of nested firewalls. A firewall (see Figure 4-1) is a system that is designed to prevent unauthorized access to the data on your servers. Such systems usually take the form of hardware routers or specialized servers, but they can be found in software-only versions as well. For optimum security, all corporate Web site servers should sit behind a firewall, as should the database servers from which the Web site pulls its information and all other system components that have a connection to external networks, such as e-mail servers.

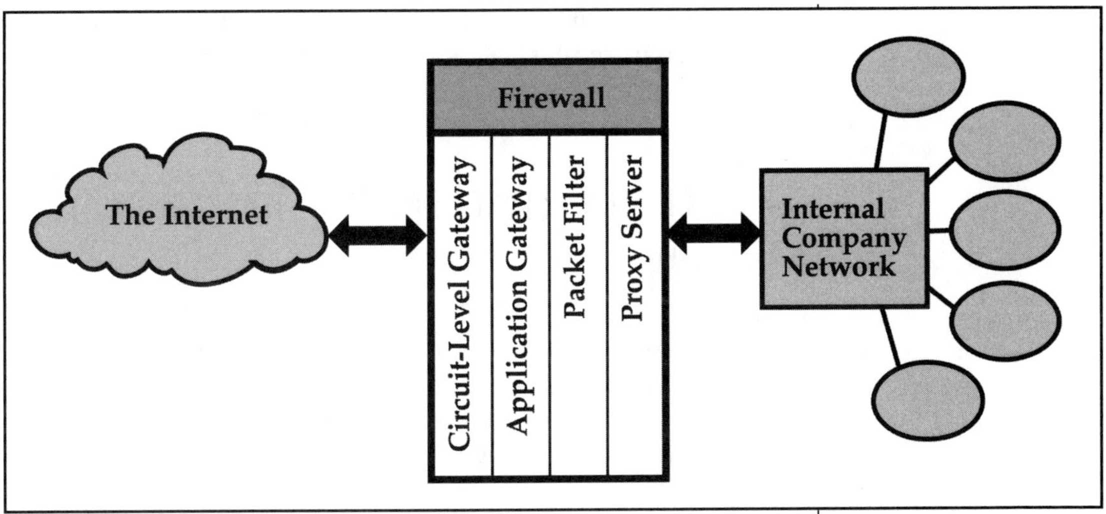

Figure 4-1. A firewall protects the internal company network from external sources using a combination of these techniques.

You can use a variety of techniques to provide firewall protection. Packet filters block network traffic based on specific IP addresses. Proxy servers function as relays between two networks and break the connection between the two pieces of hardware. Most firewalls, though, use NAT (Network Address Translation) to increase the security of the network on which the servers reside.

NAT translates a single, routable, external IP ad-dress into multiple, non-routable, internal network addresses. This hides internal addresses from the outside world. You can combine NAT with a stateful firewall that inspects each arriving packet with a carefully constructed set of rules for the firewall. This gives you a powerful tool for preventing unauthorized access to your server. Unfortunately, NAT is not a panacea since it also prevents videoconferencing and voice-over-internet-protocol (VoIP) connections from crossing into the internal network. That means Webex, Net2Phone, and other Web-conferencing and telephony solutions will not work easily within a well-protected corporate network environment. If needed, you can disable NAT in expensive firewall-and-router combinations. In its place, you can use more sophisticated anti-hacker defenses.

Another important tool for the security conscious is the antivirus software program. Also called "virus scanners," these software programs search for known viruses by looking for their special binary patterns. Databases of virus signatures are maintained by a variety of software vendors and updated regularly as new viruses are discovered. The challenge is for the corporation to ensure that its version of the antivirus software is up to date by downloading the latest virus signature files.

e-Commerce Servers and Data: Transaction Security

e-Commerce would not be possible if there were no means of ensuring the security of the transactions. Secure Web servers use cryptography (encrypted text) and special protocols, such as SSL (Secure Sockets Layer, see Figure 4-2) and its successor TLS (Transport Layer Security), to provide a secure link from the user's computer to a known server that has been identified by a certificate of security. What all this means is that secure Web sites encrypt order data from your browser before it is uploaded to the Web site. Thus, if the information you are transmitting, such as a credit card number, falls into the wrong hands it will be difficult to decipher.

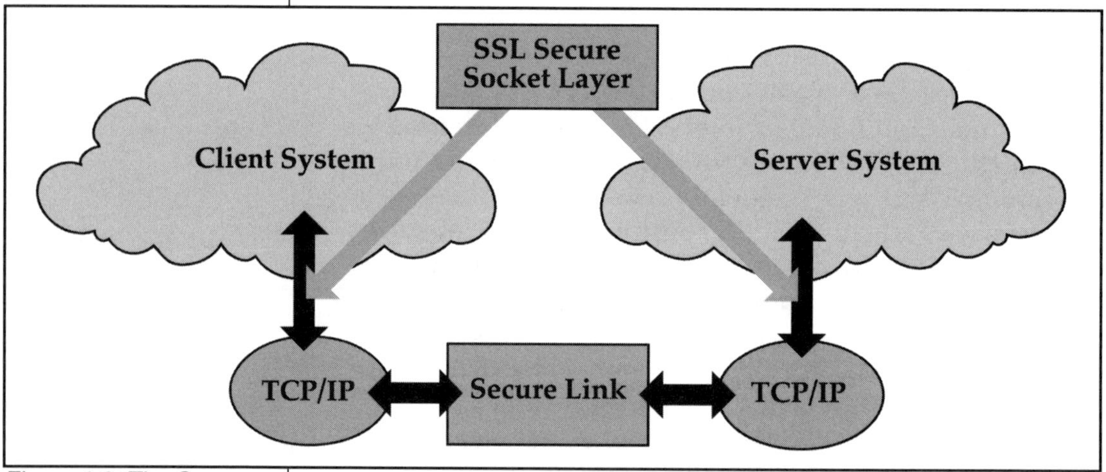

Figure 4-2. The Secure Socket Layer permits sending and receiving secure, encrypted transmissions from client to server, and from server to client.

Most Web browsers are able to check the authenticity of a Web site's certificate. After doing so, they will tell you whether the certificate presented by the server either does or does not match the expected server identification. Since version 5.0, Internet Explorer has had the ability to use 128-bit encryption. This is essentially unbreakable except by the dogged work of multiple supercomputers or the distributed computing attack of super-hackers. This means that the technology exists, and is being used, to provide a very high level of security for online transactions.

Enterprises can also protect e-mail by using cryptography that has both public and private encryption keys. PGP ("Pretty Good Privacy") is one such way of accomplishing this. The computer security firm Verisign provides simple and inexpensive encryption tools that can be integrated with Microsoft Outlook, Netscape Navigator, Eudora, and other popular e-mail clients.

Common Sense

So, with all the security technology available, why are there reports of stolen credit card numbers, for example, from Egghead.com, the computer and software vendor; Citibank; and other large institutions? Most of these cases come down to human error. Simple security procedures, for example, not storing credit card numbers on the public Internet server, would have made these thefts much more unlikely.

Virtual Private Networks

Virtual private networks have been around since the beginning of the Internet. Now, however, they are being used extensively in B2B e-commerce. A virtual private network (VPN) is a network that uses public connectivity to establish private nodes, with security, levels of permissions, and encryption. The simplest example of a VPN is the standard retail ISP. For example, I have an account at Earthlink. Earthlink operates a public Web site as well as a virtual private network for its members. This permits encrypted and secure access to data storage, FTP access to Web site back ends, and encrypted access to e-mail on the virtual private network. Companies use VPNs in the form of extranets to establish secure connections with their distributors and outside sales people.

VPN technology is improving, as is the ability to encrypt and defend these virtual networks from penetration. All B2B marketplaces are effectively VPNs, and this technology will become progressively more ubiquitous and invisible over time.

Colocation and MSSPs

When evaluating Internet services, companies are faced with a make-versus-buy decision. How will the connection to the Internet be made? Where should the Web server be located? As with many technology-based services, you have a range of options for choosing where and how you establish your Internet presence. One of the leading considerations is who is best qualified to establish a secure environment. The good news is you don't have to do everything yourself!

A popular option is to co-locate a company's Web servers with an ISP (also known as a COLO) that also hosts the Web sites of many other companies. The COLO thus provides the power supply, backup power supply, and Internet connectivity necessary to serve many high-traffic sites. In addition to their main bread and butter, namely, Internet connectivity services, COLOs also tend to provide suites of programs for clients to use. These include e-commerce and CRM applications, as well as detailed assistance with integrating applications. As you can see in the sidebar on Exodus Communications, which is one of the largest colocators, high levels of security are of utmost importance for its clients.

Alternatives to outsourcing include MSSPs (managed security service providers), which are essentially ASPs (application service providers) that specialize in security. MSSPs focus on delivering their services to small and medium-sized organizations, which often don't have the in-house IT expertise needed to guarantee that topnotch security technologies have been identified and implemented throughout their networks.

However, the economic climate is such that COLO's and ASPs are not the most stable businesses of the Internet era. The manufacturing enterprise must therefore carefully select its service provider since it could be difficult to recover if your ISP suddenly disappears, taking with it your Web server.

A Visit to Exodus Communications

The door is bulletproof and fireproof. It opens onto a very small lobby with a drive-through, bank-teller-style window, behind which is a security office. The security officer requires you to present an ID and insists that you fill out a form explaining who you are, why you are there, and who you wish to see. You must promise that you will not reveal details of the security system you are about to see.

After your identity and the purpose of your visit have been verified, your escort arrives to take you into Exodus Communications' state-of-the-art Internet Data Center. This is a top-of-the-line COLO (colocation) facility, replicated by Exodus as well as its competitors in many locations around the world. Each facility is custom designed from the ground up to be a colocation facility.

Each floor of the Exodus facility is built on a raised computer room floor, with easy-to-use ducting and wiring troughs and security provisions that include cameras, security alarms, and human guards making rounds. Each set of servers is independently located in a secured location and the building's HVAC (building environmental controls) system is capable of separately cooling or heating each location.

Like its competitors, Exodus provides redundant systems for Internet access and power as well as redundant backup power systems. It staffs its Internet Data Centers with 24x7 access to physical security experts, Internet security experts, systems administrators, and hardware specialists. As Exodus proclaims on its Web site (www.exodus.com), "The result is a physical and technical environment affording customers the reliability and flexibility necessary to outsource their mission-critical Internet operations."

Yet even Exodus has had security problems. In September 2000, the Internet Data Center in Seattle, Washington, where MSNBC, Hotmail, and MSN were colocated, fell victim to human error at the hands of its power company, Seattle City Light. Even the backup power system was fried, and it took Exodus several days to restore optimal service to their clients.

A Telling Experience

Early in 2001, when Walt had donned his persona as president of CorgiAid Inc., a charity involved in dog rescue, he received a chilling note from Anthony Templer, the CEO of HT Networks Inc., CorgiAid's ISP. Templer informed him that HT Networks' COLO had simply closed its doors and disappeared. This left Walt's servers, as well as everyone else's, in a locked warehouse with no power, no Internet connection, and a sheriff's padlock on the door for nonpayment of rent. This, Templer noted, was the cause of CorgiAid.org's offline status.

Working night and day, it took HT Networks over a week to reestablish its business, purchase new servers, program them, and reestablish DNS server identities for CorgiAid and its other customers. Since all the sites were still on the servers that were locked up in the warehouse, each customer had to reload its sites from scratch. HT Networks did a wonderful job restoring its customers. No one, however, absorbed either HT Networks' costs or those of its customers for the time and manpower it took to get back online or for

Working night and day, it took HT Networks over a week to re-establish their business, purchase new servers, program them, and re-establish DNS server identities for CorgiAid and its other customers. Since all the sites were still on the servers that were locked up in the warehouse, each customer had to reload their sites from scratch. HT Networks did a wonderful job restoring their customers, but no one covered either HT Networks' costs or the costs of their customers, in terms of the time and manpower it took to get back online, as well as the lost opportunity cost of not being offline for two weeks.

In establishing our security barriers, however, let's not forget that the ultimate goal of a network is to support communications, not prevent it.

the opportunity cost of being offline for two weeks. The moral of the story is that it is extremely important that you monitor not only the health of your Web site, but the health of your technology and delivery partners, COLOs, ISPs, ASPs, and system integrators. In addition, we highly recommend that you mirror your site on internal servers so you're prepared for all circumstances.

While this chapter has been full of horror stories, it is not intended to keep you from taking your business online or from reaping the benefits of transforming a traditional manufacturing enterprise into an extended enterprise. As with any new endeavor, there are risks and rewards. Make sure you are aware of both. The risks of extending your enterprise online are far outweighed by the benefits created by a truly integrated organization, one that communicates with its employees, its customers, and its partners, anywhere and everywhere. Putting the right technology to work behind the scenes will ensure you a positive experience.

Security and the e-Factory

by Ken Crater, President and CEO, Control.com

It's seemingly a paradox. How do you make your business transparent to customers, but opaque to invaders with less-profitable potential? For a decade or more, the thought of providing external access to critical manufacturing processes has caused more than one automation engineer to quake in fear. And yet, because what was once viewed as a maintenance convenience has now become a business imperative, these fears must be faced and conquered.

For security to be effective, you cannot view it as a component to be installed or an afterthought once your system architecture has been developed. Rather, the design of your network architecture and the overlay of services it will support must take into full consideration the security requirements of your business.

It's the System...

It starts with your networking infrastructure. How will your network be segmented and what technologies will you use to gate traffic between segments? Control and maintainability become all-important when answering these questions. Today's routers give you a comprehensive toolset for controlling data flow between network segments. This allows you to discriminate based on protocol, IP address, port number, and other attributes.

What network topology will you deploy at the device level? You may be able to save a few dollars (though the price differential is rapidly declining), but hub technology is a poor choice compared to more modern Ethernet switches both for performance and security reasons. A hub allows all network traffic to be visible at each node to which it is connected. This means that anyone breaking into a system that is connected to a hub would be able to "sniff" the network traffic on all other systems connected to that same hub. Switching technology, on the other hand, isolates the traffic to each node, creating a higher degree of protection against further compromise.

And what of your connections to the outside world? Obviously, you must take special care when designing your Internet connectivity. The standard practice today is to create a "DMZ" (demilitarized zone), that is, a network segment that is isolated from both your plant network and your Internet connection by a

(continued)

three-port firewall. Typically, only machines in this DMZ are allowed external access, and they in turn have carefully controlled access to internally sourced data. Such an architecture interposes several layers of protection between your plant and the outside world. Assuming you intend to allow a protocol such as HTTP (Hypertext Transfer Protocol, used to carry Web traffic) into your DMZ, anyone who takes advantage of a hole in your server's implementation of HTTP to compromise it would then face a second barrier: a firewall that is programmed to disallow access to your internal networks.

We're Trying to COMMUNICATE

When establishing our security barriers, let's not forget that the ultimate goal of a network is to support communications, not prevent it. Often, the best security architecture starts with a decision to *support* a certain form of communication rather than a series of negative decisions regarding which protocols to disallow. This policy might best be expressed by the maxim "everything not explicitly allowed is forbidden."

For example, SSH2 (Secure Shell) is an increasingly popular technique for communicating securely with host computers across an insecure network. It supports whatever encryption algorithm (3DES, Blowfish, etc.) you choose to encrypt your authentication sequence and subsequent communications. This prevents any illicit interception that may otherwise compromise your security system. Some SSH2 clients support tunneling by other protocols, which means file transfer and other operations are allowed to use the same encrypted channel.

For secure communications between sites, you may implement a protocol such as IPSec. This will forge a secure, encrypted communication channel that can extend across the Internet.

Once a decision has been made to support a limited range of connectivity options, all other protocols and ports can be blocked. This removes many of the holes that external threats might otherwise exploit to gain illegal access.

Are You the Weakest Link?

At the top of your security checklist should be the weakest link in the chain of defense: the human one. How many routers are there in the world today that share an easy-to-remember password like "password"? How many firewalls have had a hole poked in them "temporarily" so some test could be performed and then never see full protection restored? How many disgruntled system administrators have left a company with a book (or head) full of passwords that were never subsequently changed?

An investment of hundreds of thousands of dollars in a security infrastructure is essentially wasted unless the human factor is considered in the system design. This may be the weakest link, but it's also the hardest to fortify. A combination of policies (explicit password standards, for instance) and recurring training is necessary to achieve a reasonable degree of security. But even then, you are only as secure as your latest hire.

You must also consider the other side of the coin. When it comes time to reconfigure the router or upgrade the server, who knows what the password is? Was it the fellow who quit three months ago? Or is it the overworked system administrator with a faulty memory? Your company should put in place policies requiring the recording and secure storage of passwords to ensure that the answers to such questions are the ones you want to hear.

...It's Also the Process

This, finally, leads us to a broader conclusion: security goes beyond systems. It must be a continuing process. Not only must an initial installation be secure, but you must exercise continued vigilance when applying patches, watching system logs, and ensuring compliance with stated security policies. It is possible to attain a high level of security in a networked installation, but it is unlikely to happen by accident.

CHAPTER 5

Customer Relationship Management: Crafting Online Customer Care for the Industrial Markets

The Internet has made it possible for buyers and sellers to form even closer relationships. As a result, manufacturers are now able to learn more about their customers and deliver better products and services. Customers benefit directly by being able to spend their money wisely on products or services that more closely match their needs. In making these close relationships possible, the Internet promises to unify the entire "value chain." Closer and more frequent contact with suppliers streamlines the manufacturer's supply chain; quick and easy contact with customers allows the manufacturer to respond to customer demand and increase its "value-add."

Much has been written about streamlining the supply and demand chains and unifying the value chain, from raw material supplier to end user. Some of this is real, some is achievable in the near term, and some is just plain science fiction. One of the legitimate advances is customer relationship management (CRM). Fundamentally about sales, marketing, and customer service, CRM provides customers with increased value by enhancing the efficiency of front-office operations. In this chapter, we will look at the effect the Internet has had on the manufacturing value chain, including customer relationships. We will consider the changes that have occurred in industrial distribution over the past ten years, both on the buy side and the sell side, and the degree to which the Web has magnified these effects. We will discuss the customer service side of the equation separately, in Chapter 7.

In recent years there has been a sea change in the way customers want to relate to sellers. Today's customers are intelligent, educated people who no longer tolerate "being sold to."

Customer Relationships

It is important to remind ourselves that customer relationships didn't just appear with the advent of e-commerce. The practice of creating and maintaining relationships is as old as business itself. It might be said to *be* business itself, for without the relationships between customers and sellers there would be no business. But in recent years there has been a sea change in the way customers want to relate to sellers. Today's customers are intelligent, educated people who no longer tolerate "being sold to." The reasons are better education, changing values, and lack of time.

One of the most important reasons for the presence of this new industrial customer is that typically he or she is no longer a baby boomer. Many of today's customer are in their twenties or thirties and, as a result, have a different set of values than their parents. They are members of the post-TV generation, Internet-savvy and burned out on overt persuasion as a sales tool. Yet they buy recognized brands, eat brand-name food, and drive brands of cars that are uniquely identified as theirs.

Quite aside from generational differences, customers today increasingly have little or no time to see vendors. This means the "schmoozing" and "let's play golf" school of selling is rapidly disappearing. Today's customers want to do their own research, only calling in a salesperson to ask the final, detailed questions and place the orders. Customers want to be in control of the purchasing process. The Internet offers a tremendous opportunity for manufacturers to give them that capability

Here we run smack into the dialectic between customers and vendors where, thanks to the unparalleled access to information provided by the Internet, both now fight for control of the purchasing process. To facilitate the customer-driven purchasing process, most vendors are using increasingly sophisticated electronic tools to better understand and assist their customers through the selling process.

Data Mining

Customer relationship management is all about knowing your customers and having information about them available, from contacts and referral sources to technical preferences. Traditionally, salespeople have provided the value-add of knowing the manufacturer's customer base one customer at a time. The emergence of database technology has made it possible to collect and manage large amounts of customer data. This makes it easier for the salesperson to transfer his or her own customer information and work in concert with others to, for example, get a large order spread across several states and many offices. We can't say we're sad to say goodbye to those famous 3x5 card files!

Of course, databases aren't an offshoot of the rise of e-business. Since the 1980s, long before the development of the World Wide Web in 1994, large companies have been trying to bring together all the information they have on customers into centralized databases. Their goal is to extract exactly the information they need to make a sale. Sometimes they want to see purchasing trends, sometimes pricing trends. Sometimes they just want a reference for a specific application. Being able to find this sort of information quickly is an effective aid to good marketing and good selling.

Data mining is the least expensive form of market research because you don't have to pay someone to go out and acquire the data for you—it is already there. All you have to do is to dig through it and find what you need.

Such searching of a company's customer database is called *data mining*. It is the least expensive form of market research because you don't have to pay someone to go out and acquire the data for you—it is already there. All you have to do is to dig through it and find what you need. Special software tools have been created to make it possible to extract customer data easily. In the consumer arena, when you call a catalog (PC Zone, for example), the telesales representative has your name, phone number, address, credit history, and prior purchase information on his or her computer screen within fifteen seconds. This means the rep can offer you products that are likely to be of particular interest to you, in addition to the ones you are calling about. This electronic "upsell" is only possible because the vendor has collected relevant data about you.

Meeting Customer Needs

These data mining tools can, of course, provide a tremendous advantage to the salesperson dealing with a typical customer, whether it is in the consumer arena or the industrial marketplace. In other words, with them the salesperson can more easily meet the customer's needs. Customers have simple wants but more complicated needs. Traditional marketing focuses on manipulating those needs to increase the number and intensity of the customer's wants. It is well documented that this traditional focus has worked both in consumer marketing and in business-to-business and industrial marketing.[1] In recent years, the emphasis on "manipulation" has shifted to "transformation," but marketing is still about moving consumers from low-urgency needs to high-urgency needs. Such theorists as Jacques Werth have begun teaching that marketing and selling need not be manipulative at all. Rather, the focus of twenty-first-century marketing must be on fulfilling wants and finding customers who already want what a company has to sell.[2]

The Internet itself is inherently anti-manipulative. Every time someone has come up with a way to "push" content to a consumer online, it has failed. Why? Because most people no longer want to be manipulated. They want to be buyers, not consumers.

Toward this end, the Internet itself is inherently anti-manipulative. Every time someone has come up with a way to "push" content to a consumer online, it has failed. Why? Because most people no longer want to be manipulated. They want to be buyers, not consumers. The bottom line is that customers want good-quality products that work in their applications. In the industrial marketplace, whatever is being sold has to first and foremost work; then a manufacturer can add neat futures and a cool look. Of course, it is a huge plus if it works *and* has neat features *and* looks cool. Purchasers of technical products evaluate the "bells and whistles" after they've found the basic functionality they need. Customers want products that are useful, maintainable, and cost effective.

Industrial Marketing

Customer relationship management, which is also known as database marketing, is designed to make this kind of customer-driven selling easy. If you can access customer records, you can sort for frequency of orders, dollar amount of orders, and type of product ordered. You can, for example, tell if a customer is ordering more often than they ought to, because you can easily detect purchase patterns and determine usage requirements, and proactively send the salesman, or a tech rep, to see what is wrong *before the customer becomes angry and you lose their business.* You can also tell if a customer is beginning to take business elsewhere or if there is a downturn in the customer's business.

Let's say you are a lower-tier supplier to the automotive industry and you notice that several customers who are first-tier suppliers are buying significantly less of a particular type of component. You can do this because you are tracking orders carefully through your database. This kind of information can tell you if there is a downturn in the auto industry coming, if someone is offering an equivalent product at a much lower price, or if the automakers are phasing out the

1. Patricia Seybold, et al., *Customers.com.* Random House, 1998.
2. Jacques Werth, *High Probability Selling,* Abba Press, 1996.

device your component goes into. You can also use CRM to discover entirely unexpected things, like the fact that your failure rates are so high that you must have been blacklisted without being told!

For years, large companies have been collecting databases full of information on their customers, suppliers, and competitors. It didn't take long to figure out that this information was valuable. The availability of faster and better databases, the Internet, and better demographic analysis tools came together a few years ago in the concept of CRM. That is, you try to remember everything you ever learned about a customer and access all the information in such a way that you can use it during the sales process to increase order value, order frequency, or both. The theory is that the more you know about a customer's preferences, the better the service you can provide. Further, the more you know about a customer's preferences, the more often you can convince a customer to buy.

Industrial sales and marketing organizations have much better information about their customers on an individual basis than do most consumer marketing organizations.

The Basic Tools of e-Marketing

Once the Web took the world by storm with its easy access to online information, the Internet became a serious tool for marketing communications. Today, the Net has evolved into a sophisticated vehicle for generating and qualifying leads using such tools as content-rich corporate Web sites and finely tuned keyword advertisements on search engines. In fact, B2B online advertising reached an astonishing $1.34 billion in 2000, dwarfing even B2C e-advertising, which ended the year at $796 million.*

As with any communications program, the goals for e-marketing remain the same:

- To create name recognition within the seller's target markets
- To develop a lasting brand image for the seller and its products
- To generate high quality leads
- To close profitable sales

The Internet can help manufacturers achieve all of these goals as well as offer a uniquely global way to establish an on-going dialogue with prospects and customers.

Beyond a well-developed Web site, the basic tools of e-marketing include banner advertising, Web site and electronic newsletter sponsorships, interstitial advertising, and electronic mail.

Fully fifty percent of all electronic advertising is spent on banners (see Figure 5-1). These are the online equivalent of magazine space ads placed on Web sites. Banners have been getting a lot of bad press lately as their response rates have dropped from a high of twenty-five percent when they first appeared to today's average "click-through" rate (the number of times a Web site visitor actually clicks on the banner) of one half of one percent.

* Direct Marketing Association, "The DMA State of the Industry Report, 2000."

(continued)

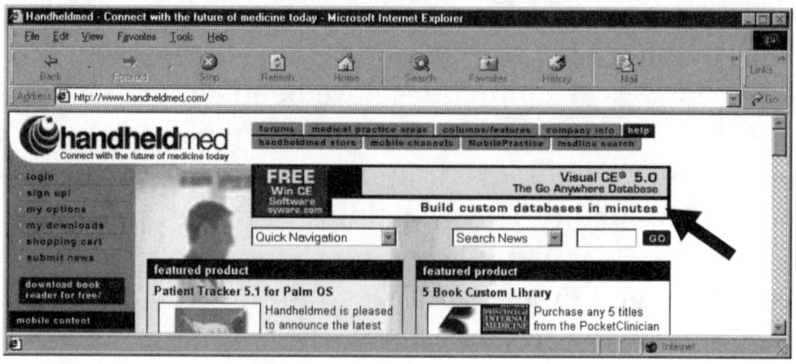

Figure 5-1. A good banner advertisement will attract attention with high impact visuals and an interesting offer.

It was inevitable that response rates would drop once the novelty of this new form of advertising wore off, but banners aren't as hopeless as the press might otherwise claim. A well-designed B2B banner advertising program can easily achieve click-through rates of three or three and one half percent, with an average cost per lead of less than $10. And as we all know, $10 per lead compares quite favorably to the $300 average cost of a sales call.

A well-designed B2B banner advertising program can easily achieve click-through rates of three or three and one half percent, with an average cost per lead of less than $10. And as we all know, $10 per lead compares quite favorably to the $300 average cost of a sales call.

A new form of banner advertisement appears on several of the major search engines (see Figure 5-2). Called "featured site links" or "sponsored advertisements," this form of advertising is a way to get your message displayed at the top of a search engine listing.

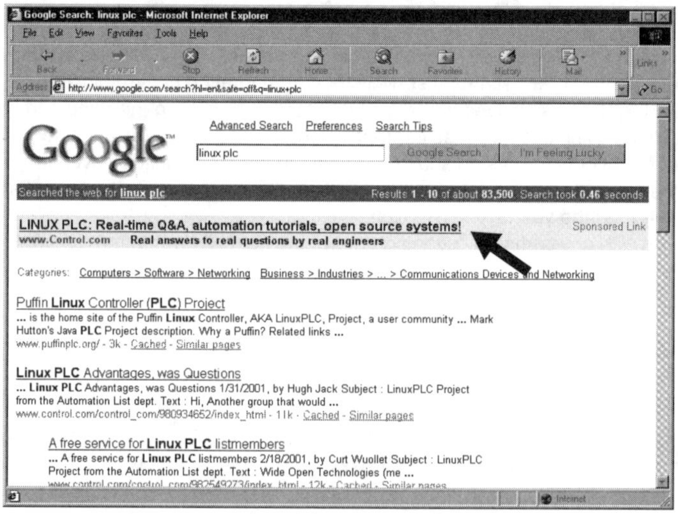

Figure 5-2. A "sponsored link" on one of the major search engines targets your advertising to only those prospects searching for a keyword or phrase related to your company.

Thirty percent of all online advertising takes the form of sponsorships. This is when a seller pays money to have their name associated with a particular Web site or third-party electronic newsletter. If your marketing program is focused on generating leads, stick to advertising in electronic newsletters. The ISA, for instance, offers a variety of e-newsletters, including ones associated with their major publications, such as *Intech*. If your company is looking to generate good will, sponsorships are an option.

(continued)

The remaining pieces of the electronic advertising puzzle involve electronic mail and interstitials. Interstitials are those annoying pop-up ads that spring up, as if out of nowhere, when you land on some Web sites. Interstitials have become quite controversial because of their "in your face" approach, but they continue to be used for about five percent of all online advertising because they generate responses.

Electronic mail has always been the real workhorse of the Internet as it is used by virtually everyone with an Internet connection. Now, e-mail, in the form of the company e-newsletter, has become a useful way to keep in touch with customers, prospects, and sales partners. These virtual communiqués, which can be sent to both in-house and relevant opt-in mail lists, are a great way to keep people informed of what's new in the way of technical content on the company Web site and new product information. Average response rates for good, information-rich e-newsletters run about five to eight percent. Adding personalization to an e-newsletter can increase response to as high as thirty eight percent.

Industrial marketing organizations can usually provide actual names and titles together with phone, fax, and e-mail numbers for individual customers. Most consumer marketing firms deal in statistics, probabilities, and profiling because they don't keep data on individual customers (although this is beginning to change with the advent of the Internet and online databases).

One-to-One Marketing

In the industrial market, it is likely that a manufacturer that keeps a central customer database has enough information to start marketing directly to individual customers. Don Peppers and Martha Rogers, who have created a consultancy based on the concept, call this "one-to-one marketing."[3]

The concept is deceptively simple. Using the tools of data mining, customer profiling, and direct mail, both postal (now called "snail mail") and e-mail, a manufacturer can selectively target an individual customer with a message so specific that it will be seen as being for that person alone. The idea is that if you can target a single person with an offer that meshes well with his or her known likes, dislikes, and purchasing habits, the chances of getting an order are significantly higher.

Peppers and Rogers offer data that indicate that targeted one-to-one marketing efforts are, in fact, many times more effective than the traditional broadcast marketing tactics of the twentieth century. Typical close rates of thirty percent to fifty percent are not uncommon; compare that with the single-digit close rates for traditional selling! The trick is to get

3. Don Peppers and Martha Rogers, *The One-to-One Manager*, Doubleday, 1999.

customers to see these focused offers as beneficial and not merely manipulative.

For a time, RealNetworks (www.realnetworks.com) suffered from a serious problem. Every week or so, its customers would get a letter from (now U. S. Senator) Maria Cantwell, who was at the time vice president of marketing. It would begin, "Dear [customer's first name]." The letter went on in a friendly and conversational fashion to convey the RealNetworks sales pitch. It was then signed "Maria." The problem is that these customers didn't know Maria Cantwell personally and were therefore pushed into cognitive dissonance instead of acceptance. When it came out, at about the same time, that RealNetworks was surreptitiously collecting information from their users' computers about their Web-surfing habits, customer relationships suffered. This *faux pas* gave Microsoft, whose Windows Media Player was lagging well behind RealNetworks' RealPlayer software, an opportunity not only to catch RealNetworks but also to put out a better product. Ultimately, Microsoft garnered a larger share of the market.

Sales Force Automation

Once a manufacturer has obtained data about its customers, it is relatively straightforward to use this information to market to them. Targeted direct marketing, by phone, mail, e-mail, and even personal sales call, is an incredibly powerful tool. If you know your customer is purchasing, for example, proximity sensors, you can use data mining to tell you what comparable customers are also buying. This allows you to target your marketing information to make sure the customer knows that you are a supplier and what the benefits of your relevant product lines are.

Using data mining to sell to a customer directly is much less straightforward. Sales Force Automation, also known as *field force automation* (FFA), has a successful implementation rate of about forty-eight percent worldwide. The rate is highest in those companies with company-paid direct sales forces. FFA is almost universally unsuccessful when it is applied by manufacturers to third parties' organizations, such as manufacturers' reps and distributors.

The Issue of Time

There are a number of reasons for FFA's fairly dismal success rate, all of which boil down to the fact that CRM systems that are designed to be used by field salespeople are time intensive. The bottom-line choice for the sales rep is, "Do I enter data into the CRM or do I make another sales call?" Since most sales reps' incentives are based on sales, the decision is not difficult; it will always involve squeezing in another sales call. Even when incentive changes are applied, this often leads to salespeople leaving for companies where they can focus on field selling rather than data collection.

In contrast, the telesales operation can use CRM to great effect. It can make a telesales and factory service operation many times more efficient, while simultaneously improving the relationship with the customer. How is it that the telesales operation can use CRM while the field sales operation has so much difficulty? A telesales operation is deskbound, with each person stationed with a telephone and a computer in front of them. The field sales rep, on the other hand, needs to interact with the customer in real time without plopping a laptop or handheld device on the desk between them.

How is it that the telesales operation can use CRM while the field sales operation has so much difficulty? A telesales operation is deskbound, with each person stationed with a telephone and a computer in front of them. The field sales rep, on the other hand, needs to interact with the customer in real time without plopping a laptop or handheld device on the desk between them.

Several CRM software vendors have begun to address such issues to tap the small to medium-sized company market. Siebel Systems (www.siebel.com), for instance, has been working in the application service provider (ASP) space, which makes it possible for smaller companies to rent the application they need, use it on someone else's server, and start an automated CRM program with little front-end cost. There are even Web-based "instant CRM" packages available, such as Salesforce.com.

The latest set of tools, and the most exciting by far, marry CRM software with personal digital assistants (PDAs), such as Palms, Handspring's Visors, and Windows CE Pocket PCs, including those with the ability to accept voice input. The main hurdle to widespread deployment is that active data entry must be done by the salesperson in the field *at the time the data is acquired.* If she has to wait until she gets back to the office or home or hotel, much of the information will either be lost, less reliable, or simply won't get done. Many CRM implementations for Field Force Automation are betting on the increasing adoption of wireless PDAs and other devices.

Historically, even getting traditional sales call reports from salespeople has been a huge struggle, winnable only by threat and intimidation. CRM reports are no different. Making it easy to produce these reports, without wasting precious selling time, is key to achieving more than the poor record that CRM has achieved to date.

Computer-mediated Customer Interactions

A considerable amount has been written about the differences between face-to-face and computer-mediated interactions of customers and sellers.[4] e-Mail messages often cause emotional discomfort when the same words spoken face to face would be inoffensive. There are many reasons for this, but the chief one is that e-mail messages lack the normal emotional cues of face-to-face interactions. On occasion, these emotional disturbances are so intense that they evoke the much-studied phenomenon known as "flame wars."

While the manufacturer may have an entire database of customer preferences, purchases, and so on, the customer may likely not perceive herself as having a relationship with the manufacturer. This alienation is exacerbated by the poor efforts of manufacturers to build community and interactivity into their Web sites. When relationship-building stops, customers tend to leave.

Flame wars have ruined friendships and destroyed relationships simply because e-mail has a limited ability to convey inflection and tone. Eighty percent of all communication is nonverbal. The same is true of Web-based communications and any other type of computer-mediated interaction. Written statements are easily misinterpreted and it is difficult to indicate irony or levity.

This was quickly recognized in the early days of computer networking. Early Internet users created a whole lexicon of emotion-conveying typography called "emoticons." Although its use is waning, the "smiley"—a colon followed by a right parenthesis—is still often seen as a way of signaling that the statement should be viewed a little more closely for emotional tone.

Let's consider the customer who is used to speaking personally with a rep or distributor. That same customer is now getting messages on her "personal order history Web site" from her e-business vendor. The relationship has now changed. While the manufacturer may have an entire database of customer preferences, purchases, and so on, the customer may likely not perceive herself as having a *relationship* with the manufacturer. This alienation is exacerbated by the poor efforts of manufacturers to build community and interactivity into their Web sites. When relationship-building stops, customers tend to leave.

4. Donna Hoffman and Thomas Novak, "Marketing in Computer-Mediated Environments: Conceptual Foundations" (Nashville: Owen Graduate School of Management, Vanderbilt University, 1994 and 1995), www2000.ogsm.vanderbilt.edu/cme.conceptual.foundations.html.

Relationship-building

In his seminal work *High Probability Selling*, Jacques Werth argues that there is a significant difference between a personal relationship between a buyer and a seller and a professional one.[5] He characterizes the latter as being based on "mutual respect and trust." Nevertheless, he states that it is possible to quickly establish a professional relationship such that a salesperson can increase productivity and raise sales closure rates to above fifty percent!

Werth's model shows the philosophical underpinnings of customer relationship management. In this paradigm, selling can be done completely manually, with no technologically based tools at all. Marrying Werth's theory of High Probability Prospecting to CRM technology tools makes the salesperson a formidable and efficient revenue generator.

Integrating CRM into the Enterprise

Many manufacturers still treat sales and marketing, and by extension, customer relationship management, as if they were outside the core functions of the company. To them, the company is fundamentally a manufacturing and logistics enterprise. As a result, CRM applications often fail because they are not allowed to access the data stored in the "silos" in manufacturing, accounting, or engineering. Figure 5-3 shows the traditional data storage model.

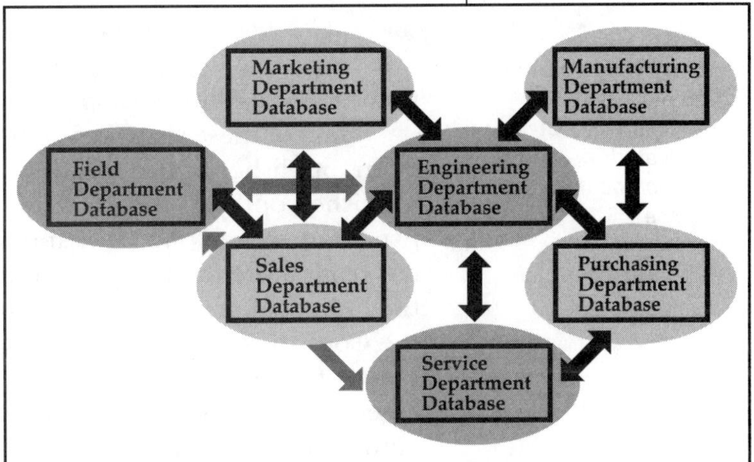

Figure 5-3. In the traditional data storage model, there are many cases where there is little contact between some parts of the company.

Note that in the traditional model there is little contact between some parts of the company and others, and often there is no information transfer. Usually, databases are different, unintegrated, and in many cases incompatible. Conversely, Figure 5-4 shows the CRM data storage model, which produces an integrated, information-rich enterprise.

5. Werth, *High Probability Selling*.

In the CRM model, all the data a company possesses is available to anyone in the company and can be drawn upon to assist a customer at any time.

Virtual Community

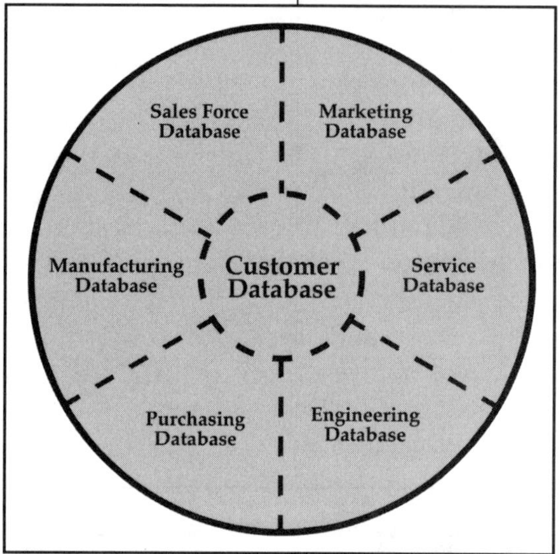

Figure 5-4. The CRM storage model produces an integrated, information-rich enterprise.

The most important concept to enter the popular consciousness because of the advent of the World Wide Web is the "virtual community." Borrowed, as were many other concepts, from the early USENET newsgroups and electronic bulletin boards, the idea of the virtual community is to electronically link a group of people who are interested in similar issues through Web chat clients, e-mail lists, or IRC or ICQ (chat clients based on the Internet's TCP/IP protocol). When brought together in these ways, people can form communities in almost the same way they would if they were in the same place geographically. The only difference is that they are located in cyberspace.

Some of these virtual communities are quite vibrant. Jim Baen, publisher of Baen Books Inc., a major science fiction publisher, operates a Web community[6] that receives between five hundred and one thousand posts (communications) per day from more than fifty different community members. In the industrial marketplace, the Automation Mailing List (see the Control.com case study on page 192) has attracted thousands of control engineers by providing a unique and valuable forum for discussing issues common to the industrial automation marketplace. e-Mail lists, such as those run by John Audette (founder of two successful Internet advertising and consulting companies, and proprietor of yet another: Adventive Inc.), can even make people like Audette quite wealthy, although fees are usually not required to belong to these lists.

There is a tremendous hunger for community among users of industrial equipment and control systems. The various user groups of the larger automation systems companies testify to this. Because of downsizing, there is sometimes

6. www.baen.com.

only one automation engineer at a given location, and he or she often has no one to turn to for questions and brainstorming. Enter the Internet and virtual communities.

Manufacturers can prosper by promoting and maintaining these virtual communities under very specific circumstances. One of them is that the "neutral ground" on which the community is "e-built" must be preserved. Companies cannot use these communities for selling, or members will flee. So, if these virtual communities are neutral, what are they good for? Ask Intel, whose cover-up of the Pentium I's mathematics flaw was broadcast through Newsgroups on the Internet. On the positive side, Apple uses virtual communities to support its legions of Mac lovers.

What can you learn by setting up and monitoring a virtual community for your products? You can start by treating it as a focus group. Uncover information about customers' applications that is difficult to get any other way. The group will provide honest answers to questions about product features and the relative merits of products' benefits. You can be a hero for providing customers with a forum in which to discuss their manufacturing problems and quickly and easily interact with their peers.

What can you learn by setting up and monitoring a virtual community for your products? You can start by treating it as a focus group. Uncover information about customers' applications that is difficult to get any other way. The group will provide honest answers to questions about product features and the relative merits of products' benefits. You can be a hero for providing customers with a forum in which to discuss their manufacturing problems and quickly and easily interact with their peers.

Creating and Managing Content

The key to a successful virtual community is providing a welcoming place for people to gather. Much like a cozy pub, library, or coffee house, a virtual community must be a place people want to visit, time and again. In the online world, the only way to accomplish this is to provide content that will draw the users to the site.

Typically this is done by creating a portal site, a virtual doorway into the world of a community. Yahoo!, Earthlink, America Online, MSN, and many others have commercial portal sites. Manufacturers also have portal sites. Schlumberger's Web site, for example, is a portal site dedicated to Wall Street analysts, shareholders, and stakeholders.

All portal sites offer the following basic features to users:

- News and information that changes daily, or even hourly; this can be general information or news specific to a particular vertical market
- Ways for members to interact, either through online chats or discussion lists
- Links to a wide array of other online sites of interest to the community
- Information on how to join the community

Web Site Development: You Get What You Pay For

As in virtually any endeavor, you get what you pay for when you develop a Web site. The following prices represent the average cost of small, medium, and large Web sites with increasingly sophisticated features as you progress from small to large. The numbers are compiled annually by *BtoB Magazine* (www.netb2b.com). The numbers show variation based on demand and the economy. For instance, the dotcom boom of 2000 pushed prices higher, but the subsequent bust of 2001 brought them down to a more reasonable level.

Average Prices for Site Development	Small Company	Midsize Company	Large Company
September 1996	$20,000	$100,000	$600,000
October 1997	25,000	80,000	275,000
June 1998	40,000	100,000	300,000
September 1999	80,000	160,000	400,000
May 2000	110,000	110,000	600,000
May 2001	65,000	125,000	250,000

Content Is King!

Content is the name of the game in creating community and in producing a Web site that customers, employees, and prospective customers will use. It is not enough to put your brochures or instruction manuals online. The more detailed and relevant the content on your Web site, the better traffic will be and the more repeat traffic you will get.

Industrial Web site content should include the following:

- Detailed product specifications
- Applications information
- Frequently asked questions (FAQs): both pre- and postsale
- Troubleshooting guides
- Drawings and wiring diagrams
- Articles and tables that would interest and help users
- Custom tools that customers and employees can use, like special-purpose calculators
- Product configurators
- Instruction manuals
- Up-to-date information about news and events

Personalized Agents

Personalized agents are one of the newest software tools you can use to help your customers learn about products and services. Born from decades of artificial intelligence research, personalized agents are robotic tools that reside on the Web and continuously search out and report on the location of online content.

The most popular personalized agent is the one bundled inside Microsoft Internet Explorer, called Synchronization Manager. When you add a Web site URL to your Favorites file, you will see a checkbox that says, "Make available offline." If you check this box, you are programming the Synchronization Manager's personalized agent to make sure that every time you go online the content of that Web site is downloaded into your Windows Temporary Internet Files folder (whose content is stored in cache memory) and synchronized with your original download of that site.

Other personalized agents exist for searching for jobs, buying cars, and looking for specific news or product information. Customers can use these agents to find what they need from the more than one billion Web pages currently online. If they prefer, they can choose to completely bypass the marketing and sales efforts of a manufacturer and focus on the core data they need.

These software agents are evolving rapidly and their popularity is growing. What this means for companies is that their Web marketing efforts must be continuous and a company's Web site must be updated at least daily to keep its information reservoir full and fresh.

The most popular personalized agent is the one bundled inside Microsoft Internet Explorer, called Synchronization Manager. When you add a Web site URL to your Favorites file, you will see a checkbox that says, "Make available offline." If you check this box, you are programming the Synchronization Manager's personalized agent to make sure that every time you go online the content of that Web site is downloaded into your Windows Temporary Internet Files folder (whose content is stored in cache memory) and synchronized with your original download of that site.

Customized Content

One of the greatest forms of one-to-one marketing is offering users the ability to personalize and customize content on a particular Web site. Most commercial portals provide this capability. There is MyYahoo! and MyMonster.com for job seekers. There is ISA OnLine Personal for industrial automation professionals. Cahners has a personalized Manufacturing Marketplace and VerticalNet's vertical portal e-magazines allow you to customize content. Customized content offers tremendous benefits for users because they can maximize their online efficiency by looking only at what is relevant for their needs. Personalization also offers tremendous possibilities for manufacturing firms that want to integrate the power of the World Wide Web throughout the manufacturing enterprise value chain, both within and outside.

Integrating the Nets: Intra, Extra, and Inter

Watlow Corporation (www.watlow.com), a manufacturer of temperature sensors, heaters, and controls, is an example of what can be done when you combine sales automation with a Web site and customized content. As a result of acquisitions and consolidations, Watlow owns five or six manufacturing facilities that have incompatible manufacturing operations software. Rather than try to make all the plants use a common system, Watlow chose to build a Web-based *intranet* (*intra-* rather than *Inter*-net since it is "inside" the corporation).

Once the intranet was in place, the company recognized that it could save thousands of dollars per year by creating a Web-based *extranet* for its salespeople and distributors. The extranet was designed to allow the sales channel to look up part numbers, check inventory status, confirm order status, and look at account status and order history for any product Watlow makes at any of the Watlow manufacturing facilities. And once the product is shipped, the Watlow distributor can seamlessly travel to the Internet to track the package at the UPS or Federal Express Web sites, using the tracking number provided by Watlow's extranet. Watlow has even arranged to incentivize its distribution force to use the Watlow extranet to place orders. It's worth the cost because if the order is placed electronically it does not need to be double-entered at each Watlow facility. This saves tremendous time and expense since many orders are for products made at more than one location.

Eventually, Watlow will open its extranet to customers, and they will probably allow customers to place orders directly for shipment from their nearest distributor's stock. Adding live online technical support via voice or video over IP would also provide Watlow with a highly efficient way to interact with its customers and distributors, while at the same time reducing the expense of customer call centers.

The Future of CRM

Gartner Group forecasts that by 2004 the supply and demand chains that move toward a collaborative CRM model will provide end users with more than thirty percent more value through streamlined pricing and higher service levels.[7] This justifies the widespread embrace of customer relationship management tools and technologies.

7. Gartner Group, *B2B CRM in a Connected World,* www.gartner.com, October 2000.

According to Gartner analyst Dan Miklovic, vice president of e-business services, the following trends are emerging in the CRM arena:

- Disintermediation (the alienation of the sales channel when a company interacts directly with its customers) is more hype than reality: Most manufacturing enterprises will quickly realize that going directly to customers is more difficult than they expected. The benefits of lower cost of sales are quickly outweighed by the manufacturer's failure to provide adequate service levels.
- Demand chain relationships are filled with uncertainty and distrust. Even though mass disintermediation is a dead concept, confusion reigns throughout the distribution channel. Such uncertainty is wreaking havoc with most CRM strategies.

As a result, most CRM efforts focus on direct channels. Most manufacturers are working to improve their own direct sales force and are leaving indirect channels to fend for themselves.

Less than fifty percent of CRM and sales force automation (SFA) system implementations are successful. Moreover, there are several unsettled lawsuits pending against CRM vendors and their integration partners. For these reasons, is important to thoroughly analyze what a CRM implementation really involves before undertaking it.

"What Makes a Good CRM Implementation?"

Less than fifty percent of CRM and sales force automation (SFA) system implementations are successful. Moreover, there are several unsettled lawsuits pending against CRM vendors and their integration partners. For these reasons, is important to thoroughly analyze what a CRM implementation really involves before undertaking it.

Start by being very clear about the goal of the implementation. It doesn't hurt to go back to the old S.M.A.R.T. system of strategic planning—make your customer relationship management system Specific, Measurable, Attainable, Relevant, and Trackable. Be clear on what you want the system to do and how you want it to work *before* you begin looking at systems and providers. For example, do you want your CRM system to be used by everyone in the company? How do you plan to enter data? How will you ensure the quality of that data? How will you clean your database? Will you have one or multiple databases? The planning that goes into this preparation stage can be more important to the eventual success of your CRM implementation than the actual implementation work itself.

The CRM field is changing dramatically. New companies, new features, and new market leaders will have surfaced by the time this book is published. Aside from the large major players, such as Siebel Systems (www.siebel.com), PeopleSoft (www.peoplesoft.com), and Baan (www.baan.com), there are a large number of newer, smaller players (in excess of one hundred companies). Some of these players are exploiting the fact that it is easier to implement CRM in a repetitive niche. That is, it is easier, more profitable, and more cost effective for a supplier to confine itself to, say, the medical information niche and install systems that are extremely similar than it is to work with doctors one day and industrial distributors the next.

ASPs and Browser-based Front Ends

The other emerging trend is the use of ASPs (application service providers). By cost-sharing the price of licensing and technical support across their entire customer base, ASPs can provide high-level CRM implementations even to companies in the $25 million per year revenue range. Siebel, for example,

One of the major reasons CRM implementations fail is the high cost of upkeep after the system is deployed. The system you choose should be maintainable by each organization or department using it and should not require dedicated internal staff, except in very large implementations.

(continued)

now offers a CRM front end for Microsoft's Great Plains division's accounting systems. Yet another new trend is the emergence of Web browser-based ASP implementations. Salesforce.com and other Web startups are able to furnish canned SFA and CRM tool sets that require no programming or local support. Some companies report considerable success using these methods. Salesforce.com claims it signed up over twenty thousand organizations in its first eight months of business, including a well-publicized partnership with IBM. No data exists, however, on the number of organizations actually using their CRM system, let alone the number of satisfied customers.

Whether the CRM system you choose is network-based or Web-based, all good systems share certain common features. These include the following:

1. *Easy to use by non-IT personnel.* This is a critical feature because one of the major reasons CRM implementations fail is the high cost of upkeep after the system is deployed. The system you choose should be maintainable by each organization or department using it and should not require dedicated internal staff, except in very large implementations. Even in large implementations, it is better to have a system that can be maintained by the people who are using it.
2. *Can easily synchronize data with personal information managers (PIMs) and contact management software, like Microsoft Outlook, Act!, and Goldmine.* Many companies, and many individual field reps, already use contact managers and handheld devices. They know how to use them and they would like to continue doing so. Some of these software products are expandable into complete CRM packages themselves, like Act! and Goldmine. Multiple Zones, the PC and Mac vendor, uses Act! coupled with an Oracle database as the CRM system for its telesales force. Implementing the CRM system will be easier if it does not require users to be completely retrained in a new software package.
3. *Easy to enter data in the field using scanners, pen- and voice-based data entry systems, and standard keyboard-based data entry methods.* The most significant barrier to implementing CRM in a predominantly field-based environment is that entering data is difficult and time-consuming. Salespeople and tech reps rarely enter their data in real time, unless it is possible to do so without unpacking the laptop and finding a modem connection.
4. *Can be used with desktop, laptop, and palmtop systems, including wireless systems that use Bluetooth and WAP protocols.* The CRM system should be capable of working on a wide variety of platforms, including the newest wireless handheld devices. This makes it possible for the field sales and technical support force to use it *in the field.* If the tech rep is standing on top of a storage tank at an oil refinery and needs to check on the availability of a control valve trim assembly, she needs to have the system available *then and there* or she will stop using it.
5. *Easy to integrate with existing databases and legacy systems; should not require you to replace existing back-office software.* More and more vendors are realizing that companies are not going to replace their existing back-office systems to realize the efficiencies and rewards of good customer management. CRM vendors should be able to easily and inexpensively adapt their systems to your existing applications infrastructure.
6. *Simple, easy-to-learn, easy-to-use reporting software so anyone can create any report without dedicated IT assistance.* The data in the database isn't worth anything if the sales force can't get at it in the form they need and at the right time. A truly enabled sales rep should be able to use the CRM system to review a customer's history or check the availability of a part while the customer is sitting across the desk from him.

(continued)

7. *Prewritten standard reports, including customer behavior prediction analysis tools.* Many marketers are not mathematicians and industrial marketers often have engineering rather than market research backgrounds. A good CRM implementation should therefore provide easy-to-use tools for analyzing customer behavior and buying habits without requiring a custom implementation for each report.
8. *Collaboration tools for workgroups.* Not only should a good CRM system enable a field rep to access any item in the database, it should also provide an integrated way to transmit important information to, from, and between the reps. A good CRM system will be the only software a field rep has to use to make proposals, check product information, provide a customer history, communicate with his peers and the home office, and enter and recall customer data.

The Problem of Disintermediation

Dramatic changes in the supply chain are behind the move toward disintermediation. Large companies ($100+ million in annual sales) have some significant advantages over smaller operations in that they tend to purchase goods and services in such impressive quantities that their supply chains are willing to do "extras" for them. These extras can take the form of maintaining special stocks of products, offering bigger discounts, even disintermediating their customers' reps and distributors.

Many large companies use company-paid sales forces rather than independent distributors so they can more easily flatten their sales organizations through attrition, reassignment, or reductions in force. Large companies also have the cash flow and credit lines to weather whatever difficulties disintermediation can cause, whether on the supply side or distribution side. Smaller enterprises rarely have this kind of financial flexibility.

Even after two decades of downsizing, large organizations have the depth of staff to be able to absorb the extra duties and responsibilities that come with disintermediation. They can also more easily absorb the additional production planning that is required when buying direct from a factory instead of relying on buffer stock stored at a distributor's location.

What are not clear are the long-term effects of both supply-side and distribution-side disintermediation. Together, we call them "the value chain." However, supply chains and distribution chains have completely different problems when it comes to disintermediation.

Loss of Advocacy

The most immediate challenge is that disintermediation produces "loss of advocacy." One of the essential support functions reps and distributors provide their customers is acting as their advocate at the factory. The more effectively a rep or distributor fulfills this advocacy function, the better the service the customer gains from the factory, especially in emergencies.

For the manufacturer, the rep or distributor also serves an important function besides sales—serving as a "triage" point for customers in need of help. Without the presence of a supportive distribution channel, the manufacturer must face all of a customer's demands, not just the ones that used to filter through the distributor or rep. This leaves the manufacturer with a significantly lower signal-to-noise ratio, which ultimately costs the company money. The customer may be lost if its legitimate requests are lost in the noise. Or the manufacturer may have to ramp up its internal customer support operation. To successfully survive the Internet era, manufacturers must make the investment needed to develop new customer advocacy tools.

For the manufacturer, the rep or distributor also serves an important function besides sales—serving as a "triage" point for customers in need of help. Without the presence of a supportive distribution channel, the manufacturer must face all of a customer's demands, not just the ones that used to filter through the distributor or rep.

The organizational structures of small ($5-$30 million in sales) and medium-sized ($30-$90 million) companies are significantly different than those of larger corporations. Small companies, for instance, lack depth. If such a manufacturer wants to handle production as well as new product development, its only manufacturing engineer may have to perform all the duties of her regular job and also work on the product development team. Salespeople in small companies regularly wear product management hats and handle marketing functions.

(continued)

Cost Shifting versus Cost Savings

Though companies may attempt disintermediation to improve margins, it is not clear that this will be its result. Cost shifting rather than cost savings is the more common result. There are as of yet no clear studies that demonstrate a net positive effect from disintermediation *in the industrial marketplace*. While both productivity and profitability gains have been reported in the electronics distribution and computer industries, it appears to cost as much to disintermediate industrial sales as it does to continue doing business as before. Margins in some industries are actually being reduced by customer-driven disintermediation.

The organizational structures of small ($5-$30 million in sales) and medium-sized ($30-$90 million) companies are significantly different than those of larger corporations. Small companies, for instance, lack depth. If such a manufacturer wants to handle production as well as new product development, its only manufacturing engineer may have to perform all the duties of her regular job and also work on the product development team. Salespeople in small companies regularly wear product management hats and handle marketing functions.

Wearing Multiple Hats

When it comes to disintermediation, the same issues apply. On the supply side, a purchasing agent who used to deal with three or four distributors now must work directly with thirty or thirty-five manufacturers. Not surprisingly, in such a scenario the purchasing agent's efficiency and work quality would quickly go downhill. Could the small company add staff to aid the purchasing agent? Yes, it could, but usually only at the risk of undermining the reason for disintermediation in the first place: the cost savings yielded by buying directly.

On the distribution side, a domestic sales manager who used to deal with thirty to fifty distributors or reps must now juggle fifteen thousand direct customers. Each of these customers expects the same level of service he or she received before disintermediation. They also expect to realize some cost savings for buying directly. To maintain service levels in this situation, the small company must sacrifice margin in order to hire more inside help. This inside help takes the place of the inside help at the distributor or rep's organization that was servicing the customer before disintermediation.

On the other hand, medium-sized companies, though they still lack many of the resources available within larger corporations, have more depth than the smaller firms. If they purchase in large enough quantities, they may be able to reap some of the benefits of disintermediation on the supply side. If they are sufficiently agile, mid-sized organizations may be able to maintain their previous levels of service on the distribution side. If they need to, they usually can afford to increase staffing levels to some extent. All in all, the odds are that a well-managed disintermediation effort by a medium-sized company can succeed, on both the supply and the customer side.

As the case study of Industrialvortex.com shows (see page 123), most industrial customers do not actually want complete disintermediation. A prominent instrumentation and controls engineer at a major Canadian refinery put it well: "I don't want to purchase from a centralized RFQ clearinghouse. I want to purchase from local distribution channels that *I* have control over and can find if I need to have recourse."

All this talk of disintermediation is stimulating some innovative concepts. For example, manufacturers are starting to pay their distributors and reps directly for the "value-added" services that used to be covered by commissions or discounts.

Testmart Inc.

Testmart was founded in 1998 by some early Web commerce players together with the principals of a brick-and-mortar business that sold used and refurbished test and measurement equipment. The original concept was to produce a content portal that would provide a nonbiased, normalized way for engineers and scientists to do side-by-side comparisons of the specifications and performance of all of the test and measurement products in the world. There would be a news feed, an applications information database, Web-based chat, and e-mail newsletters, all of which would be provided as a free service to those who registered on the site. The revenue model for the business was based on vendors paying to be included in the site's database and to receive sales leads generated by the product comparison engine.

Early on, Testmart discovered that there was great resistance to this model. Users opposed the forced registration process that was required to gain entry to the site. Vendors resisted paying to be included in the site and for sales leads. As the Internet has flourished, vendors can choose from many advertising-based portal revenue models. As a result, they are extremely selective about which sites they partner with. Testmart found the field already crowded by VerticalNet's multiple engineering sites, Thomas Publishing's Thomas Register, and GlobalSpec. These companies and others were all competing for the same limited advertising revenue from vendors.

From Portal to Virtual Distributor

So Testmart decided to change its model by transforming the company into a virtual distributor instead of a portal. It worked to acquire partnership arrangements with the vendors whose information was listed on the Test mart.com Web site, arrangements that were ultimately based on a traditional sales commission structure. Testmart also removed the requirement that end users register to access the site. And to keep their policies in line with increasingly important privacy concerns, Testmart collects no information on users of the site who do not purchase other than server statistics.

Testmart discovered early in this transformation process that less than twenty-five percent of the transactions at its Web site were entirely Web-based, end to end. Some level of human interaction was usually needed to close the sale. This required the company to staff a telesales operation. Since they had to implement a system for inbound Q&A and closure, it was a natural extension to use the resources for outbound selling as well.

Testmart also discovered it needed regional salespeople located in the territories who were capable of selling to large accounts. The selling wasn't always product oriented. Often these salespeople would be demonstrating how to use the Testmart.com Web site to find information and place orders.

Ultimately, Testmart has transformed itself into a hybrid that includes a Web-based catalog, a telesales force, a field sales force, and a strong logistics department to maintain the virtuality of their distributorship model. Testmart is less a "dotcom" and more a value-added reseller whose storefront is electronic. The company's test and measurement marketplace opened for business in January 2000. Sales grew from zero to $2 million by September 2000. Testmart expects to reach $75 million in sales in 2001.

"Testmart does much of its selling in very traditional ways," said Vice President of Marketing Julie Harrington. "The challenge for us is to convert those traditional customers to e-commerce so we can reap the cost-savings benefits the model will produce."

Because of this new model, Testmart was not immediately affected by the April 2000 "dotcom crash." The company has had a net positive cash flow since its Web store opened and in August 2000 Testmart received a third round of venture funding totaling $26 million. Testmart reported strong sales well into 2001 and improved its product line positioning by becoming an authorized distributor for Fluke test and measurement equipment in the Spring of 2001.

Testmart has invested heavily in "back-end integration" and logistics. Having lured away Mike Comstock, former senior vice president for e-commerce at DHL, the company is now working to provide an electronic commerce integration platform that is compatible with everything from uploaded spreadsheets and faxed order information to Commerce One, Ariba, and full EDI purchasing.

Exploring Sensors

Testmart planned to open a second storefront under the Testmart brand in early 2001. In studies, the company found that over fifty percent of its customers were also purchasers of sensors. So in May 2000, Testmart began investigating the potential for a sensors marketplace to complement the test and measurement market already existing at Testmart.com.

Again, the model was a virtual distributorship rather than an advertising-supported portal. Focus groups were conducted at various venues and it was determined that the store should open with flow, level, pressure, strain, temperature, and position sensors as a start. Testmart found they needed to have more than fifty vendors with over ten thousand individual products by the time the sensors storefront went live.

"One of the biggest challenges for the sensors market is to be able to produce a configuration engine that will easily adapt to any type of sensor on the market," said Don McCook, vice president of sales and alliances. "The other big challenge is fulfillment. We would have to partner, not just with vendors, but with major regional distributors as well to have access to quick-delivery inventory in the field."

In late October of 2000, Testmart elected to defer the rollout of the proposed sensors marketplace until at least some time in mid-2001. With the continuing softness of the economy, it appears this decision may stand permanently.

CHAPTER 6

Managing the Buy Side: Supply Chain Management and e-Procurement

Much has been written in the last few years about supply chain management (SCM). Some has been pure jargon; some has been incredibly enlightened but overly optimistic. Before you draw any conclusions, however, let's take a look at the nuts and bolts of the Internet-enabled supply chain and see what it's really all about.

The supply chain encompasses all of the steps that must be performed—such as procuring raw materials, parts, and component assemblies—for a company to manufacture its products. For example, the supply chain for manufacturing an ultrasonic level detector might look like that depicted in Figure 6-1.

As you can see from the figure, it isn't exactly a "chain." It's really more like a tree, but the chain concept demonstrates that at its simplest level, acquiring the materials that comprise a finished product is a two-dimensional, sequential process linking a manufacturer to its customers and suppliers.

Consumer - Retailer - Distributor - Manufacturer - Supplier

Information, products, and payments flow from one end of the chain to the other.

← Information →

← Products ←

Consumer - Retailer - Distributor - Manufacturer - Supplier

→ Payments →

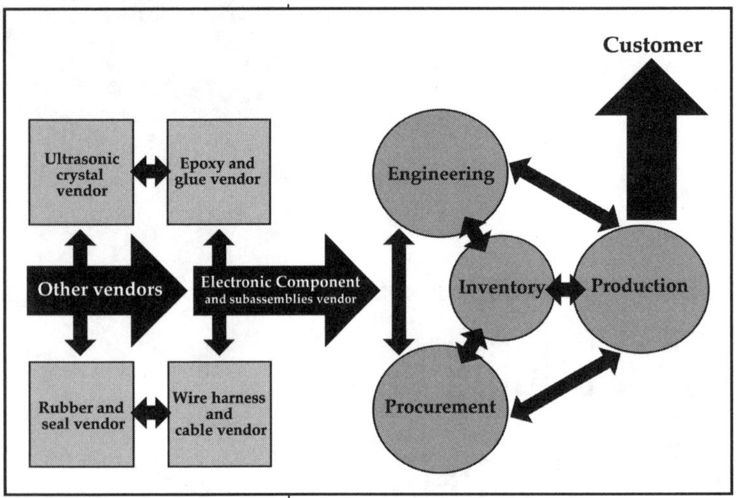

Figure 6-1. Ultrasonic level sensor supply chain.

Most companies have several supply chains, which is what gives the chain the appearance of a tree. For example, there's a main product manufacturing supply chain, an MRO (maintenance, repair, operations) chain, a facilities chain, and an IT chain. There may be more or less of each, depending on your organization's size and structure (see Figure 6-2).

It's All about Costs

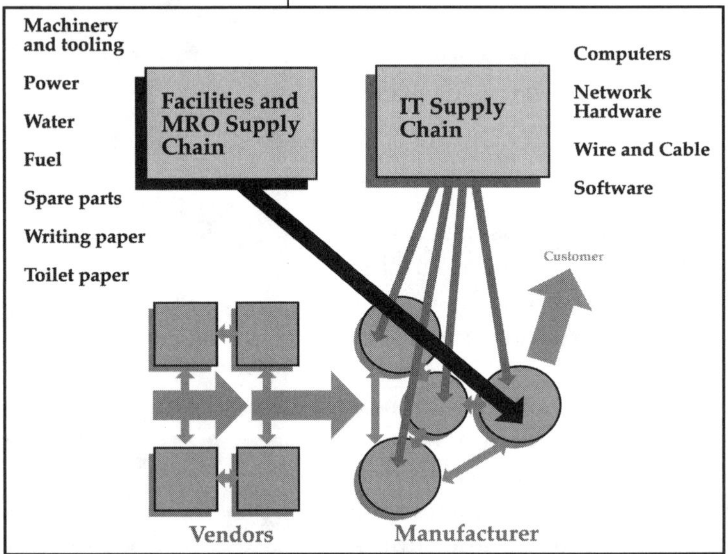

Figure 6-2. Enterprise supply chain.

What drives supply chain management is economy. Obviously, it costs more to issue ten purchase orders than it does to issue two, for example. But it's more than that. Take the cost of maintaining up-to-date files on thousands of vendors. Consider how much time this requires from engineering, purchasing, manufacturing engineering, and production control personnel. For instance, when a new product is first being designed, the engineering department must typically provide a second source for all components. If there is an alternate drawing to accommodate that second component, it must be maintained. In the words of Hugh Perry, a seasoned product design engineer at ThermoMeasureTech (formerly TN Technologies), "Changing to a different component is a tsunami that moves through the entire company, from engineering to the production line."

Engineering changes produce the same kind of turmoil as second-source suppliers. Purchasing must duplicate the work and maintain it. Manufacturing engineering, and to some extent production control, must duplicate its work and produce additional assembly drawings. Then they must provide training for line personnel on the second-source component.

The change itself can be something as simple as a capacitor with a slightly different form factor. Sometimes it's as complex as a new positioner assembly with a different bolt pattern. Customer service and technical support are impacted. Depending on the complexity of the change, marketing, technical writing, and sales can also be affected. In some cases, product sales must wait for new third-party certifications, such as UL, Factory Mutual, or CE compliance.

Clearly it doesn't take much of a change to cause major reverberations throughout a company. If there are three acceptable subassemblies, three times the overhead will be required to maintain the relevant databases, drawings, and training. Then there's the impact on inventory. If two components are acceptable, duplicate inventory is often necessary, especially for replacement parts.

__Clearly it doesn't take much of a change to cause major reverberations throughout a company. If there are three acceptable subassemblies, three times the overhead will be required to maintain the relevant databases, drawings, and training. Then there's the impact on inventory. If two components are acceptable, duplicate inventory is often necessary, especially for replacement parts.__

Companies use multiple sources of supply to prevent material shortages from causing plant outages. But it has long been recognized that the cost of using multiple sources of supply often exceeds the potential benefit.

Taken together, all these issues make it painfully clear that it is increasingly expensive to continue to design, procure, and manufacture products within the typical supply chain. As a result, companies are devoting more and more time to simplifying their supply chains. The benefits they hope to gain are the following:

- Lower cost of engineering and new product development
- Lower cost for engineering changes throughout the design life of the product
- Lower cost for procurement
- Reduced inventory requirements
- Lower cost of manufacturing by reducing need for manufacturing engineering and production control
- Lower cost of manufacturing by reducing need to train assembly workers
- Lower cost of customer service by reducing parts and spares matrices
- Improved customer satisfaction through reduced repair and spare parts costs

More impetus for simplifying the supply chain has come from just-in-time (JIT) manufacturing practices and Total Quality Management (TQM) systems. Both require a more robust and simpler supply chain than traditional production modes because of the intense maintenance necessary to keep these programs working. Generally, though, the benefits outweigh the costs as such programs can save companies millions of dollars.

Simplifying the Supply Chain

A manufacturing supply chain can be simplified in several basic ways. Some of these originate in design engineering, some in manufacturing engineering, and some in procurement. Regardless of their origin, the focus of each is on the following:

- Simplifying design
- Partnering and alliances with subassembly producers
- Just-in-time agreements with suppliers
- Supply chain aggregators
- EDI and e-business practices to reduce costs and screen potential suppliers

A manufacturer can also work toward a simpler product design, one that lowers the total number of components used in a product. Chrysler, for example, did such a good job of "matrixing" the K-car in the 1980s that they were able to become extremely profitable, even though they were using antiquated tooling in what many considered substandard factories.

Consider the design engineering function. During new product development, decisions can be made to reduce the number of acceptable suppliers for various components. These decisions are usually, but not always, made with the consent and advice of procurement, manufacturing engineering, and product marketing management. If a supplier is not on the "approved vendor" list, its products are much more difficult for an engineer to use. This reduces the number of vendors with which a company needs to coordinate.

A manufacturer can also work toward a simpler product design, one that lowers the total number of components used in a product. Chrysler, for example, did such a good job of "matrixing" the K-car in the 1980s that they were able to become extremely profitable, even though they were using antiquated tooling in what many considered substandard factories.[1] Decisions can also be made to form alliances or partner with one's suppliers. These alliances, negotiated with and through the procurement department, can include establishing agreements on inventory and giving notice of design modifications, build requirements, and shipment schedules.

1. Chrysler used an extremely small matrix of parts and subassemblies to make virtually all their car models. Everything was a K-car under the hood. As a result, the company used less than half the number of discrete parts as GM or Ford, reducing their costs tremendously.

Just-in-time supply agreements are another approach to simplifying supply chains. They typically result in lower costs, faster time to market, and reduced inventory levels, all at once. The essence of such agreements is that the supplier agrees to hold inventory so the manufacturer can procure exactly the number of parts it needs for orders on hand. In return for such an agreement, the supplier waives its usual demand that manufacturers buy large quantities of components to satisfy its product build or minimum billings requirements. The just-in-time agreement usually states that the vendor is the sole source of supply to the manufacturer. In return, the vendor must hold inventory or provide rapid turnaround on the parts covered by the agreement.

More and more suppliers are working toward the same goal—to be able to produce components on demand and to minimize the finished-goods inventory on their own shelves.

Supply Chain Aggregators

Over the past decade, many attempts have been made to use "supply chain aggregators" to simplify the supply chain process. An aggregator is a supply house (such as a cataloger) that agrees to provide most or all of the components a company uses in exchange for negotiated price and delivery terms (see Figure 6-3).

Although supply chain aggregators are able to reduce the overhead in procurement, it is not clear whether they do much else. Many aggregators have a core product line, and they then simply factor the rest of the products, components, and services that the manufacturer needs. Often, this means that the manufacturer that uses a supply chain aggregator reduces its purchasing overhead but increases other overhead, such as the complexity of design engineering and customer service. This is primarily due to the fact that the aggregator's sales and service personnel don't have the in-depth technical knowledge needed for all the products they are aggregating. By using an aggregator, the manufacturer is further removing itself from the people who know the technical details and associated production information.

Often, the manufacturer that uses a supply chain aggregator reduces its purchasing overhead but increases other overhead, such as the complexity of design engineering and customer service. This is primarily due to the fact that the aggregator's sales and service personnel don't have the in-depth technical knowledge needed for all the products they are aggregating. By using an aggregator, the manufacturer is further removing itself from the people who know the technical details and associated production information.

Many of the large electronics supply companies, such as Avnet/Marshall (www.avnet.com), Wylie/Elmar (www.wylie.com), and Digi-Key (www.digikey.com), operate as component aggregators. In the late 1990s, McJunkin Corporation acted as a supply chain aggregator for DuPont. Likewise, the Big Three automakers are funding an online supply chain aggregator (also called a vendor-backed

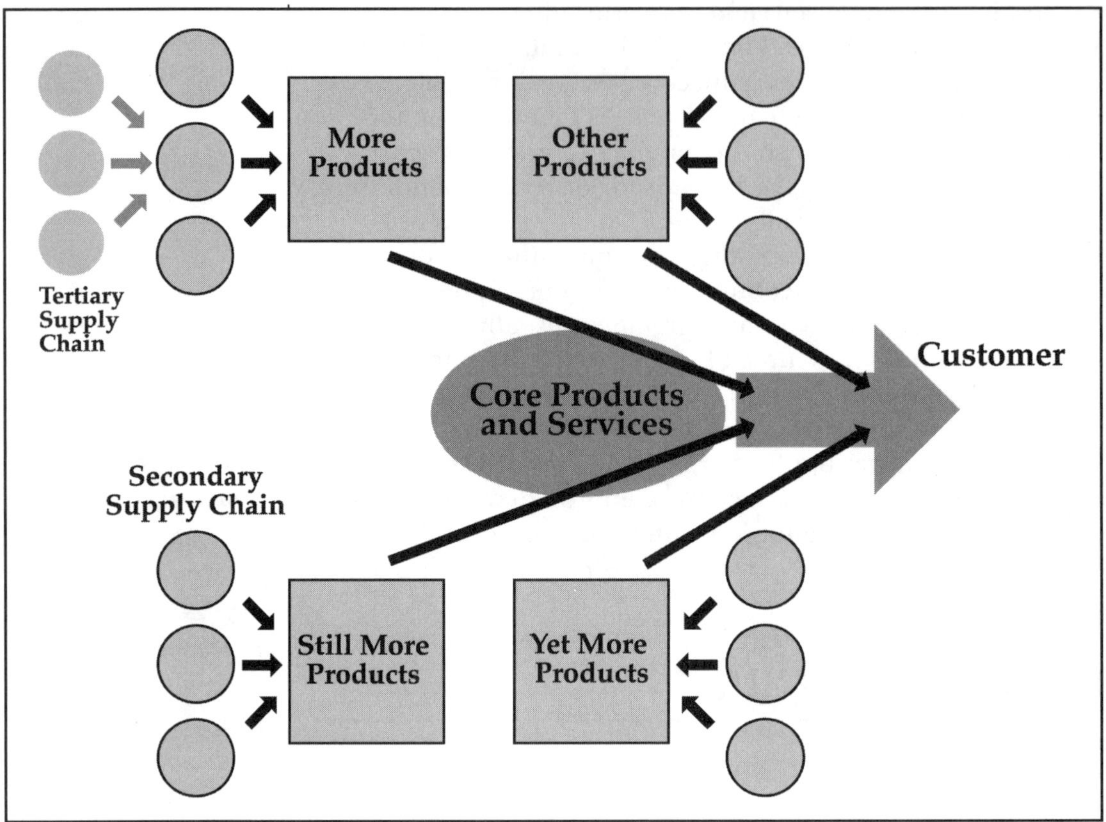

Figure 6-3. Supply chain aggregator.

e-marketplace) called Covisint (www.covisint.com), which, they expect, will cut costs dramatically. Covisint is, in fact, one of the most visible of the Web-enabled initiatives aimed at supply chain integration. The Internet and the World Wide Web have given the process of supply chain integration a remarkable boost, but it is simply the current manifestation of a long line of computer-assisted initiatives that are commonly lumped under the supply chain heading of EDI.

EDI and XML

EDI, or "electronic data interchange," has been around for some time now. It was initially pushed by the U.S. government as a way to place orders directly with large suppliers and thus cut procurement costs. Much of the cost of procurement resides in the paperwork that is needed to process purchase orders; prepare and mail invoices; and enter orders, payables, and credits into standalone accounting and production control and inventory systems. Initially, companies attempted to integrate purchasing with inventory and production control systems. The large companies began this push in the 1970s, medium-sized enterprises in the 1980s, and smaller companies in the 1990s after the emergence of the Web.

With the Web, it became possible to short-circuit the biggest obstacle to widespread use of EDI—expensive, proprietary network connections that ran between the companies and their suppliers. For EDI to gain widespread acceptance, suppliers, vendors, and customers needed a reliable way to connect their computers. Moreover, they needed to be able to communicate with each other easily and to trade data that was organized into standard formats, regardless of the internal software systems each company was running. Based on TCP/IP and standardized markup languages like HTML (HyperText Markup Language) and, today, XML (eXtensible Markup Language), the Web began to appear as the perfect vehicle for widespread standardization of electronic data interchange as early as 1995.

In the Great Web IPO Market of 1998-2000, almost any organization remotely connected with electronic commerce successfully raised large amounts of capital and went public. But reality has now set in and companies are once again focused on the basics of creating profits. Still, as more Internet tools are created to handle online transactions, it is clear that significant changes in the way companies procure components, raw materials, and even finished goods lie ahead. It is likely that the real "killer apps" of B2B will not be the technologies for electronically enabling a manufacturing enterprise, whose obvious first benefit is cost savings. Rather, enabling e-business will lead to exciting new methods of manufacturing, design, service, and sales.

It is likely that the real "killer apps" of B2B will not be the technologies for electronically enabling a manufacturing enterprise, whose obvious first benefit is cost savings. Rather, enabling e-business will lead to exciting new methods of manufacturing, design, service, and sales.

Integrating Supply Chain Functions into the Enterprise

Let's take a more detailed look at the components or functions of the enterprise supply chain and see how they are integrated into an electronically enabled business. This will show us just how great is the promise of real-time electronic information interchange. We will also begin to sense how these technological advances are radically altering business processes.

e-Procurement

The practice of procurement has already begun to change and will continue to do so until it has fully embraced the efficiencies that the world's evolving network infrastructure are making available. While processes and practices are changing rapidly, we have not yet seen the most significant effects on procurement, but they will hit their stride in the next ten years.

In the mid to late 1990s, the purchasing field embraced the World Wide Web as a search tool for information and product references. This made it possible for procurement to "shop" internationally for goods and services. In reality, the Web made it possible for purchasing agents to find other vendors for the goods and services they were already buying locally. This encouraged competitive bidding across the board, even on engineered and semiengineered products like tooling and instrumentation.

The second change was the procurement manager's ability to search the Web for surplus parts, discontinued parts, and suppliers of odd or obsolete components. With a standard browser and the Internet, a purchasing agent could find her own parts instead of having to work through a factor that always added a hefty markup to the price.

The real issue is that the overhead of running an e-market Web site is much higher than was originally anticipated. As a result, businesses whose projections were based on income from advertising and commissions are finding that their infrastructure costs are outstripping their revenue streams by a significant margin.

At this point, we saw the establishment of e-marketplaces. The jury is still out on whether these will survive to see the next decade, but purchasing agents are trying them out and working more consistently with the ones that can demonstrate a real bottom-line benefit. These e-markets act as a transparent factor, permitting buyers and sellers to find each other; they then take a commission off the top of each completed sale.

Some e-marketplaces claim they are successful, but the vast majority folded within their first year or two of existence or are in serious financial straits. The precarious position of the majority doesn't mean that the idea of a neutral marketplace isn't a good one or that the buyers and sellers are not using them. The real issue is that the overhead of running an e-market Web site is much higher than was originally anticipated. As a result, businesses whose projections were based on income from advertising and commissions are finding that their infrastructure costs are outstripping their revenue streams by a significant margin. Rockwell Automation's SourceAlliance (www.sourcealliance.com) is a perfect example. After one year, the company claimed to have reached its goal of $50 million in sales, but it was unable to secure a much needed third round of financing to support the infrastructure.

Beyond e-markets, we are seeing an increase in the number and quality of intelligent agents. These software tools enable a purchasing agent to send a "bot" (short for robot) onto the Web to look for good prices on, say, capacitors. This intelligent agent can report every morning (or when otherwise scheduled) with a list of the new prices it has found, without the purchasing agent herself having had to do anything but set the bot in motion.

All these changes are well and good, but what's really needed isn't quite ready for prime time: offering the procurement department the capability to automate the process of purchasing from its standard vendors. Commerce One (www.commerceone.com), Ariba (www.ariba.com), and other large e-commerce software developers are spending huge sums of money to produce "standardized" tools that automate repetitive purchasing functions over the Web. This is the biggest challenge that conventional, non-Web-based EDI failed to overcome. And this is where real economy will be realized.

Standardized Tools

In this scenario, the purchasing agent sets up his criteria for purchasing in the e-procurement software. He can specify price, availability, past performance index, quality index, and other basic measures, such as inventory min/max or JIT criteria. The software monitors the company's internal inventory, usage, waste, and the quantities production planning anticipates it will need. At the same time, it searches the Web for other vendors who offer the same component, while monitoring pricing on a real-time basis. When the company's usage levels reach the predetermined buy point, the software decides to whom the order will be sent and monitors the vendor's performance. The software then provides a report to the procurement specialist on all that has occurred, as shown in Figure 6-4.

The changes these standardized tools will wreak for the process of selling goods and services are quite simply astounding. For instance, it is virtually impossible to "sell" to a computer program. As a result, in the future we will see much more weight given to objective measures of a vendor's performance, more so than ever before.

How close is this model to becoming the dominant form for repetitive purchasing? Close. Such tools are already being put to use in a few large organizations since the volume of purchases involved can produce significant savings. But every company, regardless of size, should be preparing for this new purchasing paradigm—the first real killer B2B app of the new millennium. Let's take a look at some of the changes this model will bring about.

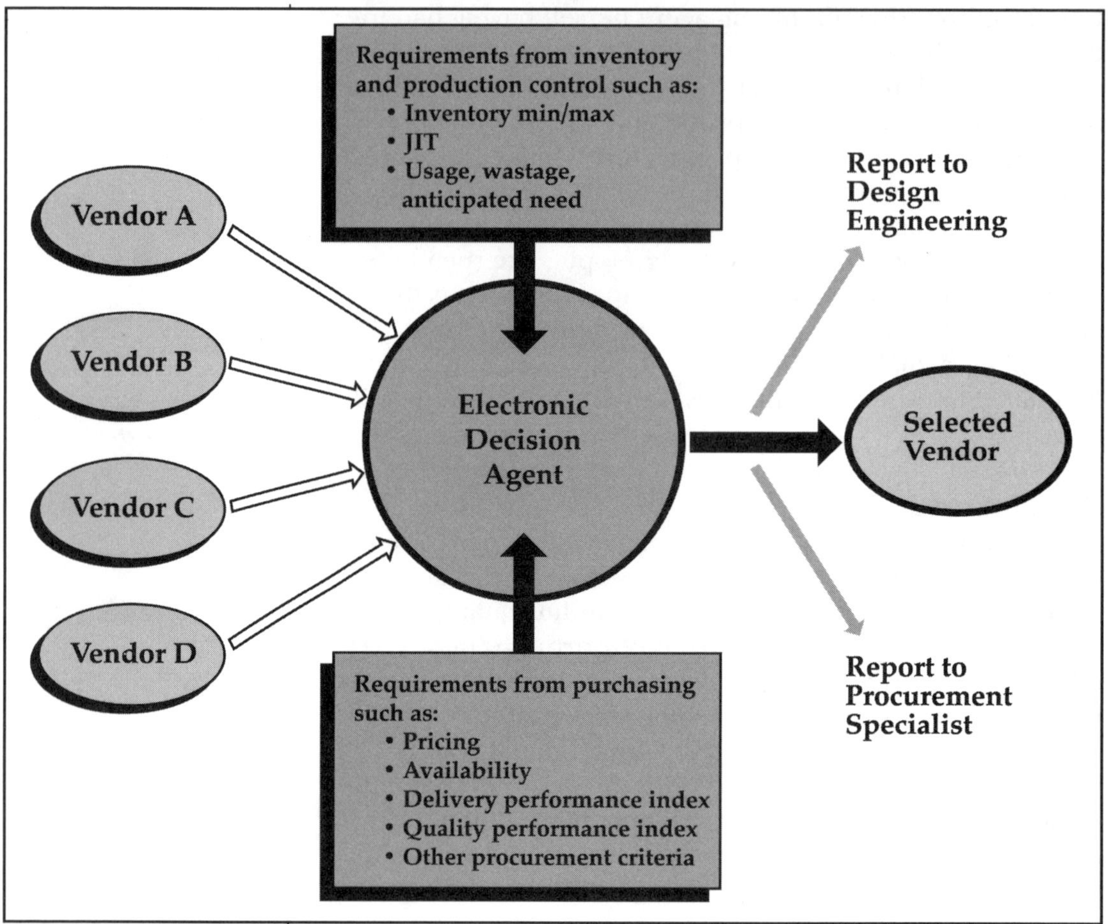

Figure 6-4. Electronically enabled procurement model.

Inventory

Web-enabled inventory functions will change as they are integrated more closely with e-procurement. It is now possible for a company to automatically generate purchase orders based on min/max settings and estimated need settings derived from production control. Soon, it will be possible to automatically generate purchase orders and even restocking orders based on pricing changes being monitored in real time. It will be possible to automatically flag soon-to-be discontinued components so they can be reviewed by engineering and purchasing, and to automatically do basic inventory management as well.

Web-enabled inventory functions are also changing the manufacturer's interface to its customers and sales force. It is becoming increasingly easy to check on a product's availability directly from the manufacturer's Web site. Soon, companies who do not implement a Web-based connection to their inventory control databases will find themselves unable to compete with those who do.

Research Bots

Purchasing agents, R&D engineers, and the general public have become accustomed to using a new research tool. It is called a research "bot," which is shorthand for a research robot software tool. A bot is an automated software program that performs a specific function over and over, just like a robot welder in a manufacturing line, or a pick-and-place robot in a semiconductor assembly plant.

Every time you enter a "search string" into a search engine and read the "hits" it has found, you are using the results of the work of one kind of bot. These specialized bots are called indexers or, more popularly, "spiders" because they "crawl the Web," indexing pages according to keywords. (What you and I do as we search online, however, is not the same as the search a bot conducts because we are conducting a manual search that we change any time we wish.)

Now, suppose you were able to set up a software tool that did your search over and over, at a timed interval such as once a day. It then returns the results to you without your having to ask or re-type your search string again and again. That tool is a bot. The most visible of type of bot, other than search engine spiders, is the price comparison bot. The most successful of these is MySimon (www.mysimon.com). While pending litigation may force the MySimon company to change its name and give up its URL, the concept is highly successful and has important applications for the manufacturing industries.

MySimon takes a search string—say, "32-inch television"—and automatically looks across the Web for items that fit your search parameters. To make MySimon work, you don't need to specify vendors in a particular industry or know all the model numbers. MySimon knows.

You can see how powerful and immediately useful the concept of bots is to the purchasing and R&D engineering functions. While MySimon is currently restricted to the B2C market, other bot companies coming online are specifically aiming for the B2B and manufacturing markets.

In this chapter, we will also talk about Zilliant, the dynamic pricing company. Zilliant's technology is an example of research bot software that is optimized to compare pricing for components and systems in the computer manufacturing industry.

Other bots are just as practical. For instance, later in this chapter, Zilliant's CEO, Deborah Vollmer Dahlke, refers to some friends who wrote a bot to find a Playstation 2 on the first day it hit the market. A little programming and they were able to find the new toy when most people couldn't. Translate this into manufacturing applications and such efficiency turns into time and cost savings for purchasing, product design and development, and inventory management and production control.

"Industrial customers send ten percent of their orders electronically today. But by 2006, distributors selling to industrial customers expect to receive thirty-seven percent of their orders online. Construction and building contractors are expected to send thirty-one percent of orders to their suppliers online by 2006, whereas by that time hospitals and schools are set to electronically dispatch fifty-one percent of their orders." Brown, Richard, "e-Procurement Faces Eclipse," Line56, www.line56.com, July 25, 2001

From a competitive standpoint, such capabilities will make it easy for you to determine if other companies in your marketspace have certain products in stock and to price your products accordingly. Such active, real-time access to inventory leads to active, dynamic pricing, as exemplified by Dell Computer and several other clients of Austin-based Zilliant Inc.

Dynamic Pricing Couples Customer Relationship Management with the Supply Chain: An Interview with Deborah Vollmer Dahlke, co-founder of Zilliant, Inc.

Deborah Vollmer Dahlke is considered by many to be an expert in the area of online customer relationship management and dynamic pricing. We interviewed her in depth to give you a more detailed look at the issues entailed in any successful supply chain and CRM integration. Most people forget that supply chain integration is about linking a chain of "vendor + customer" links. Customer relationship management is required in each of those links, and as the electronic links between manufacturers and their customers proliferate, the difference between the CRM space and the SCM space will diminish.

Question: Traditional B2B CRM applications have a less than fifty percent success rate. To help our readers beat the odds, we'd like to ask you about your CRM philosophy and how you approach the customer relationship process wherever in the supply chain you find it. Let's start with pricing. What specific information is required about a customer, and how many customers do you need to profile, to make accurate decisions about pricing strategies?

Dahlke: The best indicator of customer price response is actual behavior. In the past, conjoint analyses and price ladders were often the best way to learn about customer price behavior. The problem was that this type of pricing research only captured customer perception of how people thought they would respond to price; seldom were actual pricing decisions captured. And, when pricing research did extend to real customer price tests, the costs were high, and the time commitment required for valid responses was long—six to twelve months. Often, the resulting information was outdated before it could actually be used.

The combination of Web server technology and Zilliant's innovations in price testing and monitoring make it possible to capture and analyze real-time customer price behavior rapidly and cost efficiently. Zilliant's software tests and analyzes actual customer behavior in response to prices or price bands. Our patent-pending Zilliant Intelligent Pricing Solution (ZIPS) enables pricers and product managers to set a range of prices (price bands) around a list price and test/monitor customer action—do they take the price or not? We can also monitor other customer behavior on the site during that visit—do they choose a lower-priced product or shop around for other products? Once you have captured actual price behavior across a range of customers, you can begin to identify the demand curve or curves based on segmentation. The next step is to segment customer profiles according to similar price behavior. The profiles could be extended to include other customer-specific information, e.g., past purchase behavior, geographic or psychographic data, or other types of information.

Customer profiles can be anonymous representations of many customers that share similar characteristics or they can represent actual individual customers. The decision of how many customers are needed to be included in a grouped profile to make accurate pricing decisions really rests on what the decision is. Do you want a statistically valid sample upon which to base the decision? Or are you looking for trending information for a new product or service price?

The other aspect of this question (which you didn't ask but should consider) centers on customer privacy issues. Where privacy is a concern, it is possible to make "customer objects" with information on pricing preferences, general firmographic data, purchasing history, and other information, without the direct link to a specific customer or business. In cases where our client companies have the full consent of their customers to use personal information we build analyses and models based on identified individual customers.

* As this book was going to press, Deborah Vollmer Dahlke was in the process of transitioning out of Zilliant for other opportunities.

(continued)

Q: What makes a Web-based CRM system superior to a non-Web-based system?

Dahlke: Web-based CRM systems have several technology-based characteristics that have the potential to make them superior. However, nothing replaces the human contact and relationship aspect. Personal contact needs to be part of the equation for a superior Web-based CRM system, the same as offline systems. The technology offers opportunity for improvement, including:

- Data capture and centralized customer data warehousing.
- Internet technology—both software and hardware—enables in-depth capture and analysis of customer information, including purchase behavior, e-mail communication, individual and cluster click-stream data, sales force automation files on customers, customer registration on Web sites. The list is endless. In all honesty, few companies have figured out how to integrate and use this wealth of data effectively. But it does exist and can be used to improve the quality of customer touchpoints.
- Real-time, two-way interaction between customers, service reps, and sales teams. The Lands' End Web site is a great example. You can query the customer service rep (CSR) online about items, request a call while you are shopping, or just e-mail. What remains important is how effectively the company fulfills the promise of technology. Few things disappoint Web customers as quickly as failed responses to their e-mail queries. It is tantamount to being lost in the endless hell of automated voice mail that never asks the question you want answered.
- Better understanding of customers based on the ability to capture large amounts of information and extract key issues or trends and then respond to them.
- Neural nets and other decision-support analytics can process extremely large amounts of information to capture significant trends and concepts. For example, it may be able to identify and track key customer issues or complaints, analyze the issue, and respond more rapidly given the availability of the data and the power of analytics. I think these are emerging areas for e-CRM excellence—issue and trend tracking, analytic behavior modeling, and decision support based on the ability to capture and analyze large amounts of customer data.

Q: Is it possible to be nonmanipulative when handling customers with a CRM system? Is it desirable? How best should a company deal with customer resistance to being profiled? This is significant when the manufacturer is back in the supply chain, not directly in the retail sales path.

Dahlke: A key factor is the customers' belief and understanding of how personal or business data that is collected will be used to deliver value back to them. In the consumer realm, customer concerns with privacy will become increasingly important with customers managing their profiles more aggressively based on the expected or actual value delivered back to them. As part of this extended view of e-CRM, companies will need explicit methods (and presumably software programs) to ensure that each individual customer's data is being managed to her preferences. Included in these preferences may be varying levels of information on purchase patterns, pricing behavior, and preferences, as well as product data.

For example, I may be willing to share my purchasing and price behavior in conjunction with other personal demographic information if the company promises (and delivers) some benefit from this information that I value. Many supermarkets already capture customer purchase data and then deliver immediate value in the form of "personalized" coupons. So, you just bought Starbucks Frappuccino drinks—here is a coupon for the new flavor of Starbucks' coffee ice cream. Is this manipulative? Is it desirable? Currently, the point of view is, "If the

(continued)

customer doesn't like it there are other places to shop." In the not-so-distant future, proactive companies will take the initiative in dealing with their customers' privacy and personalization concerns. Whether it occurs by legislation or market perception, the burden will likely fall on the companies to understand how customers want their purchasing behavior managed and to clearly define what benefit or value they offer customers in exchange for access to this information.

Q: How does this philosophy fit into the movement for supply chain management?

Dahlke: On the B2B side, as customers, companies expect their suppliers to anticipate their needs and to provide clearly defined value based on an understanding of their current and future needs. It is doubtful that many businesses will object to their suppliers having a clear profile of their wants and needs—especially if suppliers get good at being able to anticipate and serve those needs as a result. Clarity of value delivered for this "personalized business information" will become a competitive advantage. Just the process of communicating how the benefits and value of keeping a business profile updated may be sufficient to influence behavior. Delivering additional value may be even more persuasive. "Thank you for participating in the Xstore's online pricing behavior test. In exchange for the information you shared we are providing you with a $10 coupon that may be used for your current purchase or a future purchase on the Xstore!"

Q: CRM systems are traditionally data-entry intensive. When faced with the choice of making another sales call or updating the CRM system, most salespeople figure it's a "no-brainer"—they make the sales call. Will the WAP, SMS [Short Message Standard Protocol], and Bluetooth initiatives make a big enough difference to enable effective use of CRM in field sales as opposed to telesales environments? If not, what will?

Dahlke: The preference for human contact versus data entry is fairly common and will persist, especially among more extroverted sales and marketing team members. However, technology and applications that make it easier to make customer calls within the wireless or office CRM system, such as automatic logging of the date/time of the call and easy access to customer records, may make recording customer contacts a more prevalent practice among sales teams. For example, I might have more incentive to edit the *DragonSpeak* transcript of my call to a customer than I would to enter it into the CRM system in the first place.

Q: Is the rapid expansion of e-CRM ASPs a blip or a real trend, and if so, why do you think so? What benefit is there to using an ASP instead of having control of the application yourself? Does this increase the rapidity with which smaller companies can use these tools?

Dahlke: e-CRM ASPs offer the same benefits (and risks) as other ASP solutions. They allow companies access to more expensive and presumably higher-value software applications at a lower cost. The value proposition for ASPs is based on the size of the company and the complexity and value of the e-CRM system offering. For example, a small or mid-size company may be able to gain advantages from using a more sophisticated e-CRM tool that has state-of-the-art analytics and data capture if it is offered via ASP. Companies will carefully weigh the risks and benefits of ASPs and service level agreements (SLAs) versus the time, expense, ease of use, and maintenance costs of installed software investments.

Q: It is clear that there is value in performing online price monitoring in volatile industries such as computers and pharmaceuticals, probably even feedstock and raw chemicals as well. Clearly, as supply chains become integrated, there has to be some electronic means of updating pricing and the tools to do it. How do you suggest pricing monitoring be performed in industries where the pricing is done off line and there are no published prices? Can it be done effectively enough to be useful in day-to-day pricing decisions?

(continued)

Dahlke: Zilliant's value proposition is to provide our clients with real customer price and value behavior and real-time, fact-based analyses and models to shape pricing strategies and actions. Most industries and companies can benefit from this type of pricing information. The degree of value may differ among industries depending on product market-price volatility. However, understanding your customers' price perception and purchase likelihood is valuable in almost any size company in any industry—from corner taco stands to global enterprises. Zilliant's solutions and technology are focused in the online world, including wireless transactions and prices/value offerings from telemarketing activities. There is an emerging hypothesis that online behavior around price/value decisions can act as a proxy for offline customer and market situations. Zilliant is testing that hypothesis but has no conclusive case evidence at this time as we are in the preliminary stages. If the hypothesis holds true, as I believe it will, online price testing can have significant value for offline and channel pricing, and promotion strategies and tactics.

Q: How would you expect Zilliant technology to "trickle down" to the basic industrial markets?

Dahkle: We don't anticipate a trickle-down. Rather, we anticipate fairly rapid adoption of technology and processes that better test, monitor, and deliver customer value in industrial and other markets.

Q: You've based some of your products on the concept of "intelligent agents." How do you see that type of technology? Will it completely transform market research? What are the limits to its effectiveness?

Dahlke: There is some evidence that intelligent agents are already transforming markets and market research. Search agents and intelligent agents are basically the same technology. It is clear that automated searches on the Internet have changed marketing. MySimon and other shopper bots make it easy to compare prices as well as other price/value information such as availability and shipping charges. Automated agents are routinely launched against auctions as "bidding robots" that carry instructions to increase bids by specific amounts with a ceiling delineated.

A group of my friends programmed an agent to watch several sites for availability of Playstation 2 during Christmas 2000. They were successful at capturing several systems that appeared on Wal-Mart's site for only a few seconds. In the not-too-distant future we will see purchasing agents routinely launch intelligent agents with instructions to visit specific exchanges or sites to negotiate and purchase routine items. It is doubtful that agents will engage in intense negotiations where skill is a factor anytime soon. That requires another level of intelligence that most bots are incapable of today.

Q: Will CRM and intelligent pricing ultimately make the consumer-user-buyer more liberated, with more choices, or will it circumscribe the customer's ability to control the selling process? This is a very large concern in the Supply Chain Management space, where vendors are concerned that supply chain integration will remove their ability to compete if they are not already the preferred vendor. How will dynamic pricing affect this?

Dahlke: Intelligent pricing provides the seller and the buyer with more information in the exchange of value. Most economists believe that more information improves the purchasing process for all the participants, including the industries. No process that limits or circumscribes buyers will be successful in the long run. Innovation will always provide options and opportunities for improvement—in pricing and in e-CRM.

Q: Amazon recently made waves with per-customer pricing strategies and RealNetworks has had real problems with cognitive dissonance on their privacy policies. Is this endemic to CRM? Is it overuse of a new technology or is it overreaction on the customers' parts?

(continued)

Dahlke: Companies can make mistakes with technology, as with anything else. In the Amazon case there was a combination of effects. The price tests being conducted covered very broad price bands that caused customer concern and feelings of unfair treatment. However, only a handful complained at first. The media caught the issue and made it into a news event. In the long run, I don't believe it damaged the potential for dynamic pricing. It did increase awareness of the potential for testing prices and also heightened the awareness of the need to establish value in association with dynamic price differences.

Sales Channels

The capacity to do open inventory search and dynamic pricing will have an effect not only on more efficient inventory management but also on sales and marketing. As intelligent agents or "bots" become ubiquitous, well-informed customers, be they purchasing agents or design engineers, will have less need to speak directly with a salesperson.

For sales organizations, the question is how to maintain relevance and provide value-added services when a customer may know more about a specific product than the salesperson.

For sales organizations, the question is how to maintain relevance and provide value-added services when a customer may know more about a specific product than the salesperson. Sales organizations are adjusting by moving to a more customer-directed sales strategy. Different approaches are used for informed, knowledgeable customers who want to be in control of the procurement process, on the one hand, and customers who expect the salesperson to serve as their knowledge base, on the other.

But the knowledge base has a potential problem. Many companies in the engineering sector are downsizing both management and operations people. While this contributes to lean, economical manufacturing operations, it also leads to the disappearance of whole volumes of institutional knowledge. As a result, we are seeing the birth of a new type of value-added reseller (VAR), one that operates in parallel with the usual channels, namely, company-employed direct salespeople, manufacturers' representatives, and stocking distributors. This new type of VAR adds a heavy service component to its offerings, thus helping to fill the void caused by lean manufacturing organizations.

For example, in the petrochemical industries, companies have regularly downsized their plant engineering departments, with the biggest staff reductions occurring in such niche areas as process analytical specialists. These are the people who provide the systems that test products in the manufacturing process to ensure that quality controls are met. To replace these people, new organizations have sprung up that handle sales, engineering, systems installation, and maintenance. This trend is an example of the fusion of the distributor and the systems integrator into a

VAR channel that did not previously exist in the manufacturing sector.

Production and Production Planning

The fine line that divides production planning and procurement is growing thinner. And with the automation that the Internet promises to deliver to many of these processes, it will soon only be a question of procurement absorbing many of production's functions or production simply doing much of its own procurement. In other words, it may be a case of six of one, half a dozen of the other.

The fine line that divides production planning and procurement is growing thinner. And with the automation that the Internet promises to deliver to many of these processes, it will soon only be a question of procurement absorbing many of production's functions or production simply doing much of its own procurement. In other words, it may be a case of six of one, half a dozen of the other.

Using product forecasts, production control now has the ability to automate the procurement of all but the most critical components. Unfortunately, the nightmare of the "running change" still exists. Real-time feedback from vendors that indicates the availability of components and subassemblies will help alleviate this nightmare. It will permit dynamic scheduling changes that take into account both macro and micro variations in the product market.

For example, it may be possible to know that a competitor's lead-time on a product is long, while your company can build its product quickly. In some cases, however, a company may be engaged in a sales campaign to gain market share from a competitor that has long lead times. In this case, it might make sense to schedule a longer production run of these products to provide quick delivery to the sales function. The ability to make such running changes easily, automatically, and electronically will finally retire the spectacle of production control assistants careening about the production floor picking up and manually replacing outdated work orders. Automating such processes makes a company more flexible and lowers costs.

Shipping and Distribution

Electronically enabled logistics and distribution have already produced tremendous change, although much is transparent to those outside the company. We still see warehouses and trucks and we still see deliveries being made and inventory carried. But what we don't see is the truly radical change going on inside the warehouse, inside the truck, within the duties of the delivery person, and within the distributors who carry the inventory.

Think about how much things have changed. It is now possible to place an order online, pay for it online, get order confirmation and shipment tracking numbers online, check delivery information online, and receive proof of delivery online. It's easy to take the next step because networked systems are already being put in place to automate the manufacturing process and automatically provide detailed information. More than half the value of installing information feedback systems is the value of the information itself, not the cost savings associated with automating a manual task. The cost savings may pay for the automation expense, but the value of the information is pure profit.

More than half the value of installing information feedback systems is the value of the information itself, not the cost savings associated with automating a manual task. The cost savings may pay for the automation expense, but the value of the information is pure profit.

It is now possible for a distributor to carry less inventory, provide the same or better delivery, and devote available resources to customer service rather than warehousing. It is also now possible for a company with "pick, pack, and ship" products to outsource warehousing and logistics to companies like FedEx, Airborne, DHL, UPS, even the U.S. Postal Service. With this kind of system in place, the manufacturer may be able to do away with local distribution and pocket the savings. Of course, much of these savings can be eaten up in higher customer service overhead, overhead that is no longer being handled by the distributor. While the Internet makes it possible to do more with fewer distributors, it also makes it possible to handle customer service directly on the Web, with human intervention required only to handle special accounts or problems. This returns some of the overhead back to the profit line.

Accounting and Finance

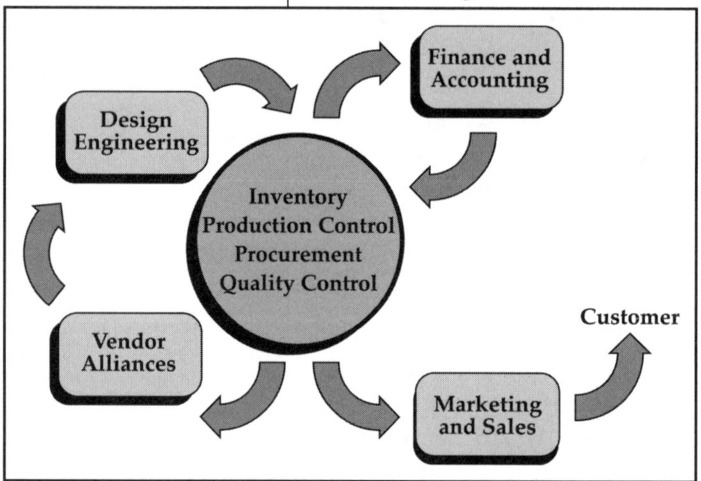

Figure 6-5. The electronically enabled enterprise.

Electronically enabling the manufacturing enterprise will also have a startling effect on compartmentalization and departmentalization throughout the organization. This means that just as production and procurement saw a blending of functions and responsibility, so too will accounting and finance. As Figure 6-5 shows, the Internet is rapidly leading to a more efficient, cross-connected enterprise.

Many of the processes now performed manually in the manufacturing enterprise will soon be automated. Process parameters will be established manually, while actual processing will be handled electronically. This is analogous to the development of SCADA (supervisory control and data acquisition) systems in the water and power distribution industries in the 1970s and of distributed process control in the chemical and hydrocarbon processing industries in the 1980s. In the same way that SCADA systems replaced the "valve jockey" and the "meter reader" with automatic data transfer systems, electronic business tools will replace the procurement specialist who manually checks prices every morning. This is making it possible for plant operations and management to work on higher-level issues, such as quality management and removing constraints.

The Internet is enabling us to progress from microbenefits, such as direct cost savings, to macrobenefits, such as indirect cost savings, and even to metabenefits, such as the synergies between production and pricing that we have seen in process control.

Customer Service

Such synergies are already cropping up in the area of customer service. Here we are seeing the benefits, both to customers and to the enterprise, of intelligent agents, data mining, and other customer relationship management solutions. Such e-business tools have had a profound effect on customer service, raising the bar in such areas as response time and personalization.

A company that can offer real-time access to inventory, order history, order placement, and order tracking can expect increased loyalty from its customers. But before long all companies will provide such services as a matter of course, just as all companies today use fax machines and e-mail. It will not be enough to just offer such communication options; a company must continually refine such services as part of its continuum of value-added services.

A company that can offer real-time access to inventory, order history, order placement, and order tracking can expect increased loyalty from its customers. But before long all companies will provide such services as a matter of course, just as all companies today use fax machines and e-mail. It will not be enough to just offer such communication options; a company must continually refine such services as part of its continuum of value-added services.

Marketing

The emphasis in this chapter has been on the use of intelligent electronic systems to empower manufacturing enterprises. For marketing and sales, the flip side is that the twenty-first century will require that the manufacturer interface *to* electronically enabled customers. This means techniques will need to be developed to make it possible for marketing to be involved earlier with customer projects, to create alliances with customers rather than simply selling to them, and to more deeply integrate the customer into the marketing and sales experience.

Customers who have electronically enabled procurement, production, and manufacturing systems are customers who are in control of the buying process. In the twentieth century, marketing and selling techniques permitted the seller to control the process by using proprietary information, applications expertise, and unpublished pricing. The tools and techniques of the electronically enabled enterprise make such approaches impossible since customers, by necessity, will possess the same amount of informa-

Customers who have electronically enabled procurement, production, and manufacturing systems are customers who are in control of the buying process.

tion as the sales force about pricing, availability, and competitive products.

When the descendants of such primitive intelligent procurement agents as MySimon (see sidebar on Research Bots, page 109) are finally able to automate the information-gathering portion of the procurement process, sales reps and distributors will find themselves refocusing their efforts on the value-added they bring to the process.

SCM Today and Tomorrow

The promise is great for the process of electronic data interchange—the nonproprietary, Internet-based version. But while many companies are working on a variety of Web-based supply chain management systems, the reality is that much of the work requires a large amount of custom coding, so much so that it will be a while before there are cost-effective, turnkey solutions that can simply be bought and installed.

Other challenges lie ahead for supply chain management, including:

- Creating interoperable and interchangeable systems, such as Rosetta.net
- Creating standards for item descriptors and syntax, including XML, ANSI/ISA-95.00.01-2000, and UDDI
- Creating a standard for SCM databases
- Producing turnkey, object-oriented tools and systems for enterprises

Currently, SCM technology is too expensive to serve the thousands of companies who are generating in the range of $30 million to $300 million in annual revenues. Instead, SCM is applied almost exclusively by the $1 billion plus enterprises. For Internet-enabled SCM to succeed and become ubiquitous, it must be translatable into products and services for mid-size companies. For the next few years, the SCM wave will move swiftly. It will be a true balancing act for the enterprise to learn how to match its IT investments with its ability to quickly and easily interface with both customers and suppliers.

The Future of Supply Chain Management

There is no such thing as "business as usual" in the twenty-first-century enterprise. We are looking at dramatic changes in the way business is organized and conducted. The managed supply chain will collide with the managed customer relationship. The result of the collision will be an integrated approach to enterprise operations from the front office (marketing, sales, customer support) to the back door (procurement, supply chain integration, production control, shipping, and distribution). The successful manufacturer will integrate intelligent agents throughout its operations and use them to seek out customers as well as good deals on components. And it will all happen globally, 24x7 (twenty-four hours a day, seven days a week).

Not long from now, supply chain management will cease to be a separate discipline. In fact, as the Internet becomes more ubiquitous throughout the enterprise, there may be no more separate disciplines. Enterprise management is clearly coalescing into one single, integrated whole.

Not long from now, supply chain management will cease to be a separate discipline. In fact, as the Internet becomes more ubiquitous throughout the enterprise, there may be no more separate disciplines. Enterprise management is clearly coalescing into one single, integrated whole.

Transforming the Manufacturing Enterprise

To transform a company into a true multidomain B2B extended enterprise is a complex exercise. Interwave Technology Inc.'s senior consultant, Charlie Gifford, shared with us a twenty-point checklist that will help transform a traditional manufacturing organization into a fully extended enterprise with a completely integrated supply chain and full B2B data interchange capability.

To begin, a company must develop, integrate, and maintain metrics vertically through the single domain of the enterprise and then horizontally across the extended domains of the company's supplier and customer base. At each step of the process, the company must achieve buy-in, develop detailed requirements and business cases, and apply the metrics to determine how much progress has been made toward an extended enterprise.

The following details the key points that a company must include in any such checklist for measuring its progress toward becoming a true e-business.

1. Achieve upper management buy-in. Before anything else, upper management must be educated in the theory of manufacturing enterprise integration. The board and the executive committee must be committed to, and clearly understand, the process of building an infrastructure for cultural change in the corporation. They must also understand the process that must be followed to become a high-response extended enterprise.
2. Perform a technology maturity and gap analysis. The team responsible for enterprise transformation has to perform two assessments before it can proceed to implementation:
   - A. Feasibility Assessment
     - i. Identify the main corporate strategy for the extended enterprise.
     - ii. Identify the gap between the technologies and infrastructure that the company currently uses and those that are necessary for an extended enterprise.
     - iii. Identify the current and future role of the internal supply chain.
     - iv. Identify the gap between the technologies and infrastructure that are currently available and those that are required to fully integrate the enterprise supply chain.
     - v. Identify all of the corporate views of the current and future state of the integrated supply chain and clearly define their business roles in the extended enterprise model.
     - vi. Determine the costs/benefits of switching to an integrated supply chain model for the operation of the extended enterprise.
   - B. Requirements Assessment
     - i. Develop a plan to scale the IT systems, enterprise-wide, and produce a reliability analysis.

(continued)

   ii. Determine the probable failure points and data storage and bandwidth load requirements.
   iii. Perform an assessment of the skill sets of the in-house team and determine how to remedy any gaps identified.

Once the preliminary work is done and upper management has been fully educated about and bought into the project, the enterprise team can begin to look at specific tactics.

3. Identify the business drivers and quantify the return on investment (ROI) for the enterprise's integrated supply chain strategy.
   A. Perform a supply chain optimization study to determine the business drivers and choke points for the supply chain, both vertically in the enterprise itself and horizontally across the supplier and customer base of the enterprise.
   B. Perform a plant-level feasibility assessment to identify the requirements for supply chain functionality at the plant level in the real world.
   C. Create a preliminary data interchange model and requirements.
   D. Do a real-world check by determining the ROI of each function in the proposed supply chain integration.
   E. Make sure that the manufacturing arm of the enterprise is fully aligned with the supply chain business strategy and correct any outstanding issues.
4. Develop an extended enterprise and supply chain strategy and model that is designed to be "low maintenance."
   A. Write the integrated supply chain strategy and implementation plan, complete with phases, goals, and objectives, and the metrics for each.
   B. Write the manufacturing execution infrastructure (MEI) plan. This process includes creating a project steering team, writing an operations manual for the steering team, developing communications criteria, creating a change management system, and establishing a version control system for both software and manuals. This plan should also include complete details about the document management system, including the documentation standards needed to fulfill the project goals. Last, the MEI plan should include complete details about the management of the project, including chain of command, responsibilities, authority, and reporting.

Once the plan is written, the enterprise team can move to the physical pilot stage. At this point, every enterprise will need to do some experimenting to determine the most effective way to integrate the supply chain and become an extended enterprise.

5. Develop a manufacturing execution infrastructure (MEI) plan during the pilot phase that will respond appropriately to the dynamic change/knowledge demands of a multidomain environment. In the pilot phase, the earlier enterprisewide planning cycle takes concrete form in a single plant environment. So, we will see some of the same assessments and reports, but this time they will be focused on a single plant entity in the enterprise.
   A. Perform a plant-level feasibility and readiness assessment.
   B. Perform a plant-level functional requirements assessment that includes detailed requirements for the entire plant's operations and the business case and ROI for each requirement.
   C. Determine the pilot project's functionality and the phases, goals, and objectives that are necessary to achieve the desired functionality.
   D. Draft a plant manufacturing execution system (MES) design specification.
   E. Determine the software selection criteria for integrating the supply chain.

(continued)

F. Apply the criteria and produce a software selection process report.
G. Design and deploy the pilot manufacturing execution system.
H. Benchmark the deployment and produce a performance assessment.
I. Perform a real-world analysis of what was learned and earned by the pilot project, and determine what changes are required to the pilot MES model.

Once the enterprise team has worked through several iterations of pilot projects, it can begin to design an enterprisewide support system.

6. Design a global support system (GSS) for manufacturing.
   A. Draft a GSS design specification.
   B. Draft a complete set of infrastructure requirements that will function globally throughout the enterprise.
   C. Draft an implementation plan and deploy the support system within the enterprise.
7. Finalize the development of data interchange requirements between the MES, ERP, SCM, CRM, and logistics and fulfillment applications by using the data acquired during the pilot testing of plantwide systems.
8. Develop an extended strategy for an enterprise application interface (EAI) layer that may incorporate the company's suppliers and customers into a scalable XML data set and schema. Finalize the specifications for the integrated supply chain (ISC) data interchange design.
9. Figure out all the ways the project can fail, with a failure effort and mode analysis (FEMA).

Now, in the period before the enterprisewide implementation, the enterprise team must go back to the pilot process. This time, however, they are focusing on a multiplant extended enterprise pilot project.

10. Extend the manufacturing execution infrastructure (MEI), global support system (GSS), enterprise application interface (EAI) layer, and the failure effort and mode analysis (FEMA) into a multiplant, multidomain pilot project.
    A. Perform a multiplant plant-level feasibility and readiness assessment.
    B. Identify detailed requirements for all the plants in the project, including business case and ROI for each requirement.
    C. Draft a multiplant design specification.
    D. Draft an integrated supply chain specification for all the plants in the project.
    E. Draft and deploy the multiplant MES (manufacturing execution system).
    F. Draft and deploy the multi-plant ISC (intregrated supply chain).
    G. Benchmark and measure the performance of the system "as designed."

Finally, it becomes possible for the enterprise team to develop the corporate deployment plan and to fully implement the systems they have piloted throughout the enterprise. To do this successfully, they will need to add some new items to the "checklist."

11. Develop rapid application deployment (RAD) teams to respond as quickly as possible to the need for enterprisewide deployments of new required applications.
12. Develop and publish the corporate deployment plan.
13. The company's education and training requirements are critical to the success of an enterprisewide deployment of new integrated systems. A complete enterprise training plan must be created that includes end-user training, super-user training, MEI and GSS training, and the relevant documentation for each training program.

(continued)

14. The multiplant pilot MEI (manufacturing execution infrastructure) and GSS (global support system) must be scaled to corporatewide levels. This project phase must include determining enterprisewide data storage and load requirements, a complete scaling and reliability plan, and benchmark criteria for the scale-up.
15. Actually deploy the MEI and GSS into the enterprise and begin widespread use.

Finally, after it has transformed the enterprise's infrastructure and technology, the enterprise team can look outward to the supplier and customer base of the company.

16. Select preferred suppliers and customers that can be integrated into the enterprise system's EAI/XML schema. Work with each supplier and customer to implement the data-interchange process and train personnel in using the system.
17. Negotiate the key process indicators (KPI) with supply chain partners by creating a stakeholders' steering team that consists of members of the internal and external supply chain partners. Publish the extended ISC data interchange model and requirements through this team. Have this team identify detailed requirements on a plant-by-plant basis, including a business case and ROI for each requirement.
18. This steering team should develop the specifications for the final pilot test: a pilot extended enterprise (EE), including ISC pilot implementation specifications and a performance assessment of the system "as designed."
19. Develop and track B2B performance metrics using the pilot extended enterprise (EE) as a model, and create benchmarks and tracking criteria for a full implementation of the extended enterprise.
20. Develop a relationship management department to administer and expand the B2B extended enterprise.

CASE HISTORY

IndustrialVortex.com: A Cautionary Tale

In the October 27, 2000, issue of *ARC News Summary*, there is a short article about Ventro (www.ventro.com) and the apparent failure of B2B neutral marketplaces. ARC's vice president, John Moore, says, "It's hardly surprising that neutral netmarkets are finding that transaction fees are simply not enough to keep their marketplaces viable." Ventro, which introduced Chemdex as one of the first neutral netmarkets, has finally admitted that its business model is flawed. "There are simply not enough customers available in any of the markets Ventro has developed to maintain a self-supporting business that is based primarily on transaction volume," he concluded.

But long before the demise of Chemdex, there were signs that the neutral marketplace model was not ready for prime time. One of the biggest signs was the collapse in July 2000 of IndustrialVortex.com.

IndustrialVortex.com was started by some of the biggest names in industrial automation, including Chuck Steinberger, IndustrialVortex's chairman and founder (and formerly of Intellution, Groupe Schneider, Square D, and Modicon), and Dave Smith, CEO of IndustrialVortex and a former vice president of marketing at Wonderware. Together, they amassed a truly awesome array of talented people who worked to build the concept of a neutral marketplace that was devoted specifically to the industrial automation and sensors marketplace IndustrialVortex.com went live in March 2000, with an initial capitalization of nearly $2 million. They closed their doors a mere four months later, in July 2000. In an interview with the authors, former chairman Steinberger stated that IndustrialVortex was forced to close because it failed to attract a second round of financing.

The company's original model was to focus on supplying industrial engineers with the tools, technology, and information to drive commerce—that is, reduce the time it takes to select and purchase a product. According to Steinberger, the tools and software required to do this are expensive and sophisticated, and IndustrialVortex was not able to raise enough capital to build the infrastructure it needed.

One of IndustrialVortex's biggest problems was a sense among manufacturers and distributors that the company's goal was to disintermediate local distributors. "You can't disintermediate anyone," Steinberger says. "When you intersect the supply chain, you have to provide added value to the distributors instead of disintermediating them. You cannot radically change behavior that much."

Another problem the company faced was the announcement by several of the major automation consumers, including DuPont, Dow Chemical, and the Big Three automakers, that they had their own supply chain integration plans. These consortia froze the market and made new capital acquisition difficult. Partly because of the entry of large corporations into the private and public exchange marketplace and partly because of the 2000 dotcom stock market crash, public exchanges like IndustrialVortex.com grew out of favor with venture capitalists. "Thus," Steinberger notes, "we became an unfundable business model."

In the weeks before its demise, the company continued to face down technical hurdles, such as properly matching RFQs with local vendors. In addition, IndustrialVortex attempted to reinvent itself as an application service provider (ASP) for local distributors. Management felt that the best way to add value to the supply chain was to provide a way for many local distributors to increase their reach and broaden their ability to sell online. Since the software to sell industrial products online (configurators, three-level catalogs, etc.) is extremely expensive, IndustrialVortex believed they could spread their costs over many distributors and thus survive. To this end, they developed a core B2B exchange product called IBOS, which they hoped would be marketable in their new role as an ASP. But it was not to be.

Steinberger still insists that the future of the supply chain is in integrated supply. The more successful Internet-based businesses will leverage technology to support existing trading relationships. "The dynamics have changed such that only consortium-owned exchanges will survive and there will only be one or two in each vertical market. We still have a few years to wait to see them become profitable," Steinberger concludes.

CHAPTER 7

Customer Service and Support

In Chapter 4, we discussed the growth of customer relationship management from both a systems and a marketing perspective. Now we want to look at the customer service portion of the e-business enterprise in more detail. We'll first look at customer service itself and then at how it is integrated into the e-business enterprise.

Changing Nature of Customer Service

Everyone has a "customer service horror story." They are a part of the times and it hardly seems possible today to find someone who has not been treated badly by a company. It's true that we hear more about bad customer service because people are more motivated to talk about their bad experiences than their good ones. But bad customer service is not just a B2C problem. It occurs in B2B interactions as often as in consumer marketing situations. So why is there bad customer service?

Probably the most common reason is a misunderstanding of the important role service plays in the growth of a business. Another reason is that customer service costs money, while good customer service costs even more. Many companies attempt to run their customer service by splitting them into two departments: one for handling presales questions and answers, which is a pure cost center, and a service department for handling post sale questions, repairs, and upgrades, which is expected to be a profit center.

Another reason we see so much bad customer service is because few companies actively and continuously train their customer service employees. Often they are poorly prepared and drawn from entry-level candidates at low pay levels. In many cases, companies expect their distribution channels and customers to bear a disproportionate share of the burden of service, but without providing the training or information flow that would make that expectation achievable.

"Many organizations resort to one of the 'big three excuses' to solve customer problems:" 1) "Well, we've never heard of that before." 2) "You must be doing something wrong." 3) "It is <insert name of other vendor>'s fault. Check with them."

Unfortunately, this means that many companies meet the public with their lowest-paid, most poorly prepared employees. These customer service agents are not trained negotiators. Generally, they hear from customers who have a problem and then they must resolve the problem. As a March 2001 article in *Fast Company* noted, "many organizations resort to one of the 'big three excuses' to solve customer problems"[1]:

- "Well, we've never heard of that before."
- "You must be doing something wrong."
- "It is <insert name of other vendor>'s fault. Check with them."

All of which end up enraging the customer. Not exactly the best way to build loyalty!

Measuring Customer Satisfaction

Even companies that have good customer service training programs and talented supervisors find that they are still perceived as providing poor service. The University of Michigan School of Business has been maintaining the American Customer Satisfaction Index (ACSI) since 1994. The ACSI is heavily weighted toward those sectors of the American economy that are interfacing with household consumers, but anecdotal evidence suggests that the data maps well to the industrial consumer, too. The ACSI directly measures customer satisfaction in thirty to forty percent of the U.S. economy and is quite representative of the economy as a whole. This index, which can be found at www.bus.umich.edu/research/nqrc/acsi.html, shows that customer satisfaction in the United States has declined since the benchmark year of 1994. In the fourth quarter of 1994, the CSI for the entire U.S. economy stood at 74.2, the highest it has ever been. In 1997, it dipped to an all-time record low of 70.7 and has been in the 72s ever since. For example, the fourth quarter of 2000 showed a decline from the previous quarter to 72.6.

1. "But Wait, You Promised ...," *Fast Company*, March 1, 2001.

The easy explanation for this decline is that customers' expectations have risen throughout the 1990s, and in fact there is some evidence to support this. But it is equally plausible that before the 1990s, the average state of customer service was so bad that, in the words of the poet Richard Farina, it was a case of "Been down so long, it looks like up to me." It is not unlikely then, that both explanations are true. Customer service has improved, but not enough to keep pace with expectations.

Defining Good Service

It is easy to know bad customer service when we see it. It is much harder to define good customer service. We know that the customer is not always right. We know that the customer who is the most demanding may actually offer the company the least value. Good customer service is, therefore, the level of service that most aligns with the image the company wants to project and the customers' expectations based on that image.

About twenty years ago, a small instrumentation company based in Germany and the Netherlands took on a project to provide a large quantity of magnetic flowmeters for a uranium enrichment facility in South Africa. By accident, the meters (close to fifty) were shipped with the wrong electrode material. When the uranium and fluorine compound flowed through the flowmeters it literally dissolved them. The plant engineering and maintenance staff called Germany and the service rep told them, "Not to worry—we fix," and hung up the phone. The South African officials were pleasantly surprised less than forty-eight hours later, when the service rep flew in with over fifty sets of electrodes and the field calibration gear necessary to produce new calibration constants in the field. The level of service this company provided more than met the customer's satisfaction. One of the engineers later told us that he'd never seen customer service that matched it, and from then on his company (then the largest engineering construction firm in the world) was going to specify that company's flowmeters on every job they did.

A company that claims a high level of customer satisfaction raises the expectations of its customers. Thus, customer service is an exercise in brand management as well as a practical means of satisfying the customer. Brand management is done well when its message is carried throughout the enterprise. So, too, customer service is done well when it becomes an integral part of all departments. When all the employees of a company consider themselves customer service agents, the company's customer satisfaction index will rise, and dramatically so.

Why this is important becomes clear when we compare the data on the cost of customer acquisition with the data on the cost of customer retention. Studies performed since 1990 indicate that the cost of acquiring a new customer is between four and ten times the cost of retaining an existing one. This means that, in the end, even though the cost of providing good customer service is not small, it is much smaller than the cost of losing a valued customer.

Personal Interaction

Since 1990, most large companies have installed elaborate call centers, replete with voice mail, call waiting, and customer databases, to help them assist and personalize service to customers. They became victims of the law of unintended consequences, however, because these technologies put more distance between the companies and their customers. The very methods that were intended to make it possible to help more people for the same amount of money wound up alienating customers and making them feel less cared for.

Another important issue that is often overlooked in the debate over customer service is the effect of a computer-mediated customer service environment. Since 1990, most large companies have installed elaborate call centers, replete with voice mail, call waiting, and customer databases, to help them assist and personalize service to customers. They became victims of the law of unintended consequences, however, because these technologies put more distance between the companies and their customers. The very methods that were intended to make it possible to help more people for the same amount of money wound up alienating customers and making them feel less cared for.

Smaller companies continue to have real people answer telephones instead of automated voice response systems. Not surprisingly, they tend to have better CSI ratings. Of course, there will always be companies, large and small, that provide bad customer service and have low CSI ratings. Size isn't the real issue. One of the biggest concerns for customer service in the new millennium is that the issue of customer response to computer-mediated interactions will become critical as more companies move their customer service operations onto Web-based platforms.

To a customer, the most memorable interaction with a company is in person. It is not just the face-to-face aspect that makes the interaction satisfying, however. It is the interaction with a specific person or type of person. This person is a company employee with whom the customer has formed a relationship of mutual trust and respect. That is, she is a person the customer trusts to know the answer and to fairly represent both the company and the customer in all discussions. Failing that, an interaction with a customer service person who is a similar to the customer is the next most personal interaction. Still, this must be a person with whom the customer can form a relationship of mutual trust and respect. It is possible to mutually create such a relationship in a matter of minutes, either in person or on the phone. All it takes is a well-intentioned staff with good training.

Since the telephone is already one step removed from the intimacy of a personal visit, how can a company make up for that? Studies have shown that customers who wait for long times on the phone are more likely to have a lower CSI than customers who get to speak to a service rep quickly. Thus, one of the ways to make up for the diminished "personalization" of the interaction is to respond to the customer swiftly. This does not mean letting a voice response system answer the phone, but having a competent, well-trained CSR (customer service representative) speaking with the customer quickly.

One of the ways to make up for the diminished "personalization" of the interaction is to respond to the customer swiftly. This does not mean letting a voice response system answer the phone, but having a competent, well-trained CSR (customer service representative) speaking with the customer quickly.

Moving Service Online

How does this translate to the online world? Beginning in the late 1990s and to increasing extent today, many companies moved their customer service response operations to the Internet. How many times have you called a company and had to listen to the "answers to most common problems can be found on our Web site; check there first before you call"? Recently, companies have taken to disabling the "0" button, which used to be a way to bypass the voicemail jail. As a result, it's no longer possible to short-circuit the recorded triage. Another new trend is the two to three minute "commercial" while waiting. Most of these commercials sing the praises of using the automated (read "cheaper") customer service features of the Web site.

Unfortunately, customers who take their service problems to the Internet face an even more impersonal system than a large-volume call center. Companies tell their customers that answers can be found faster by clicking through the company Web site. In today's technically savvy society where the convenience of self-service is valued, finding answers on the Web certainly appeals to us as a viable option. However, most corporate Web sites are so poorly designed and so lacking in useful content that customers discover that the answer to their question is rarely in the knowledge base or FAQs (frequently asked questions). This means the customer has to call the company back, listen to the voicemail triage (which often runs longer than two minutes), and try to speak with a knowledgeable person...again. For customers who have already spent hours searching the Web site and have reached their boiling point, the result is a significantly reduced Customer Satisfaction Index (CSI).

The customer comes away from the experience knowing that the company's statements ("You can easily find what you need in our FAQ") aren't always true. The result is cognitive dissonance and the more it snowballs the more the company's brand is damaged.

The customer comes away from the experience knowing that the company's statements ("You can easily find what you need in our FAQ") aren't always true. The result is cognitive dissonance and the more it snowballs the more the company's brand is damaged. Remember our definition of good customer service: "The level of service that most aligns with the image the company wants to project and the customers' expectations based on that image." The wise manufacturer will make it easy on customers with simple touches like adding a voice mail option that says, "If you visited the Web site and did not find what you need, press 1", then link the customer to a live customer service rep.

Given all the psychological baggage and brand-image maintenance issues we've discussed, it is not surprising that the ACSI has declined since 1994. Yet, with the cost of a personal sales or service call exceeding $300 million in the industrial marketplace today, it is simply not feasible to have all service interactions "in person." A well-designed customer support center on the Web will do wonders for your CSI, but it must be done in an organized and thorough manner. Include a detailed set of FAQs grouped by product, technology, and use, as well as multiple options for contacting the company should further questions arise.

Waiting on Hold

Many of the problems of customer service are systemic. For example, consider a call center example presented in *Fast Company* article that we discussed earlier. A customer waits on hold twenty minutes, while the CSR is waiting on hold for another CSR to answer his question. I experienced this phenomenon myself while writing this chapter. I was on hold with United Airlines for five minutes. When the reservation agent returned, she explained that nobody knew how to process my request (I wanted to use a travel voucher to buy a new ticket). They had to find someone who knew the answer before she could get back to me.

The reasons for such problems are varied. Sometimes it's an issue of staffing. It is unlikely that any company can staff its customer service department well enough to handle "rush hour" call volumes to avoid degradation of service. Sometimes the issue is the availability of the right information. Other times it's a concatenation of many things that just don't work right. Often, the failure of one system will lead to catastrophic failure of other systems. In a customer service situation this means that something that started as a disaster will only get worse.

e-Enabling Customer Service

In the e-business marketplace enabled enterprises are changing and for good reason. To begin with, the enterprise itself is demanding more of the customer service function. Not only do the field sales and telemarketing departments need customer service support, but the entire enterprise needs access to customer information on orders, fulfillment, shipping, and technical requests.

Real-time data on in-field failures can be immediately useful to several groups within the enterprise. Manufacturing can use customer data to spot quality control failures. Incoming inspection can use the data to gauge the accuracy of its procedures. Purchasing can use the data to see if it is ordering the right components. Manufacturing engineering can use the data to start designing a better process. Design engineering can use the same information to look for possible product improvements and potential directions for next-generation designs.

Real-time data on pricing, sales, and lost business can be immediately useful in the same ways. Information on product use can also be useful since innovative applications can help drive new product development. The key here is that customer service data needs to be completely available to any part of the extended enterprise that can benefit from it.

Real-time data on pricing, sales, and lost business can be immediately useful in the same ways. Information on product use can also be useful since innovative applications can help drive new product development. The key here is that customer service data needs to be completely available to any part of the extended enterprise that can benefit from it.

Changing Information Flow

Let's look at the traditional organization of a typical customer service department, as shown in Figure 7-1. Note its essentially hierarchical nature and how information flow mirrors the hierarchy.

As in the *Fast Company* call center example, the customer spends much of her time on hold because the CSR himself is on hold with a higher-up, who, in turn, is on hold with another department. Information from the customer flows up the hierarchy and to other departments, then down the other departments' hierarchical chains. Return information follows the same path in reverse. The biggest customer service problems result implicitly from this hierarchical information architecture. As customer decisions are pushed higher up the management chain, they become more removed from the customer and are more likely to yield a result that is unresponsive to the customer's issues—two of which are the need for fast decisions and quick support!

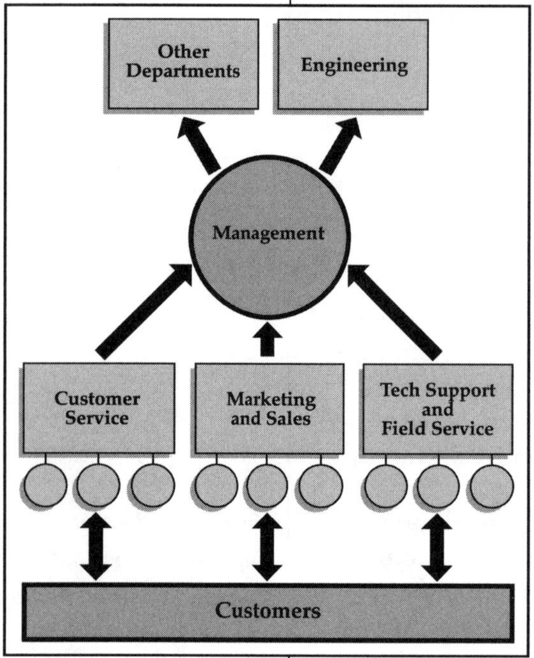

Figure 7-1. The traditional customer service organization is based on a hierarchical flow of information

Now, suppose that all of the information the company possesses about any subject, any product, or customer were deposited in a huge companywide database that any unit of the company could have real-time access to (see Figure 7-2). In this model, access is accompanied by empowerment (what IT types call "permissions"). That is, if a customer service rep has access to information about the customer's problems, he should also have permission to act entirely on his own initiative. This flattens the company's decision-making process, making the company more nimble and responsive to the customer. But it also makes the entire organization more responsive to the customer's problem. As soon as the CSR enters data describing the customer's problem, intelligent agents report the problem to the most relevant company units.

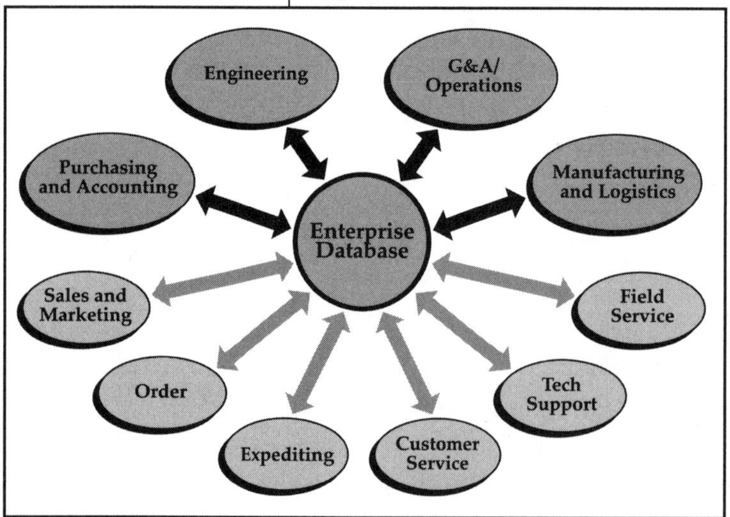

Figure 7-2. The corporate "knowledge-base" empowers the organization to better respond to its customers.

The development of such an "empowered" organization has been limited by the software and labor costs involved in implementing and maintaining a centralized database. Only companies with revenues greater than $50 million can usually afford such an IT investment. As a result, many small organizations are eyeing with great interest the growth in P2P (peer-to-peer) computing and wireless P2P. As Figure 7-3 shows, the same flat company structure that facilitates instantaneous and complete access to information throughout the company can also be used in a P2P environment. This means that companies can be organized and trained as clusters of mostly autonomous units linked in real time to information, rules, procedures, and company culture by interpenetrating and interoperating databases. Field service and field sales personnel are now empowered to interact with the company database online. This empowered approach will go a long way toward improving the meager forty percent success rate of today's CRM applications.

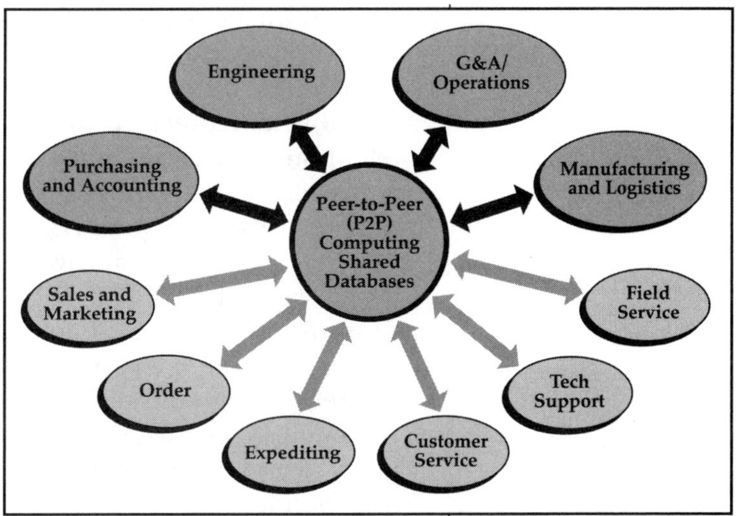

Figure 7-3. The same flat structure that facilitates access to information throughout the company can be used in a P2P environment.

Why does all this matter? Why make such a big deal about customer service? Because it has been shown, again and again, that it is up to ten times less expensive to keep and sell to an existing customer than to find and sell to a new customer. It is therefore critical to a company's profitability to maximize its ability to keep and increase the loyalty of its customer base. Doing so will increase profits derived from sales to existing customers and produce the enormous benefit of positive word-of-mouth referrals. This, in turn, makes sales to new customers easier and less costly. Word of mouth also reinforces the company brand and image in a way that cannot be paid for. It all comes down to providing quality products backed by quality service.

e-Customer Service Technologies

It should be clear by now how a call center can be improved by empowering it with a variety of e-business functions. Providing customer service employees with access to the entire corporate database and granting them permission to do what is reasonable to answer customer questions will solve most of complaints voiced about customer support today. Such empowerment also enhances the quality of the service itself by allowing companies to effectively deal with customers' ever-increasing demands for better service.

As discussed in the sidebar, The Tools of Online Customer Service (see page 135), new Web-based technologies have begun to ameliorate the "distancing" or alienation computer-mediated environments create. Already, many companies have rolled out "live help" functions that enable a customer to use a simple chat client to speak with a CSR.

The advantage of this to the customer is clear—the chance to speak to a real, live human being in near real time. For the manufacturer's benefit, a single CSR can now easily "multiplex" between customers. Instead of handling just one telephone conversation at a time, a CSR can handle as many as five chat conversations over the Internet without any degradation in service quality...assuming the CSR is a good typist! Chat clients will improve and bandwidth will become more readily available through broadband connections. This will make it possible for CSRs to present themselves to the customer over the Web in the guise of static "avatars" (either photographs or cartoons) or through an actual streaming video image of the CSR. Customers will be able to put a face to the company and develop a longer lasting rapport with it.

Even more importantly, the integration of database technology over intranet and extranet networks has enabled customer service to participate more actively throughout the enterprise. In such an integrated system, customer service can now help production scheduling personnel expedite purchases and manage inventory control, assist purchasing staff in dealing with vendor problems, and keep sales and marketing informed of potential customer issues, like long lead times and possible defects.

More than anything else, however, integrating all of a company's information resources makes it possible for customers to "self-serve" directly on the Web. No more phone tag or sitting on hold waiting for a CSR. Now customers can go online, check stock, check order status, and, if the company distribution channels are set up, place or modify an order directly online.

More than anything else, however, integrating all of a company's information resources makes it possible for customers to "self-serve" directly on the Web. No more phone tag or sitting on hold waiting for a CSR. Now customers can go online, check stock, check order status, and, if the company distribution channels are set up, place or modify an order directly online. The Web also enables vendors to provide either automatic or semiautomatic updates by going online and posting order status, price changes, delivery updates, and other customer service tidbits further up the value chain.

The bottom-line issue, of course, is the integration of the company's information and how to interface the company's information with the reams of data from outside the company. New technologies such as XML and UDDI as well as new software "translators" are being introduced almost daily. In Chapter 9, we will take a closer look at these new technologies.

Customer Service in the Extended Enterprise

It is doubtful that the enterprise of the future will even have a separate function known as "customer service." The reason it exists today is to provide a forum or platform through which to gather and disseminate the information that is required to serve the customer. In the architecture of the extended enterprise shown in Figure 7-4, you can see that any unit of the company should be able to perform its own portion of the customer service function. Customer service will become like the sea in which the fish swim…invisible yet necessary and ubiquitous.

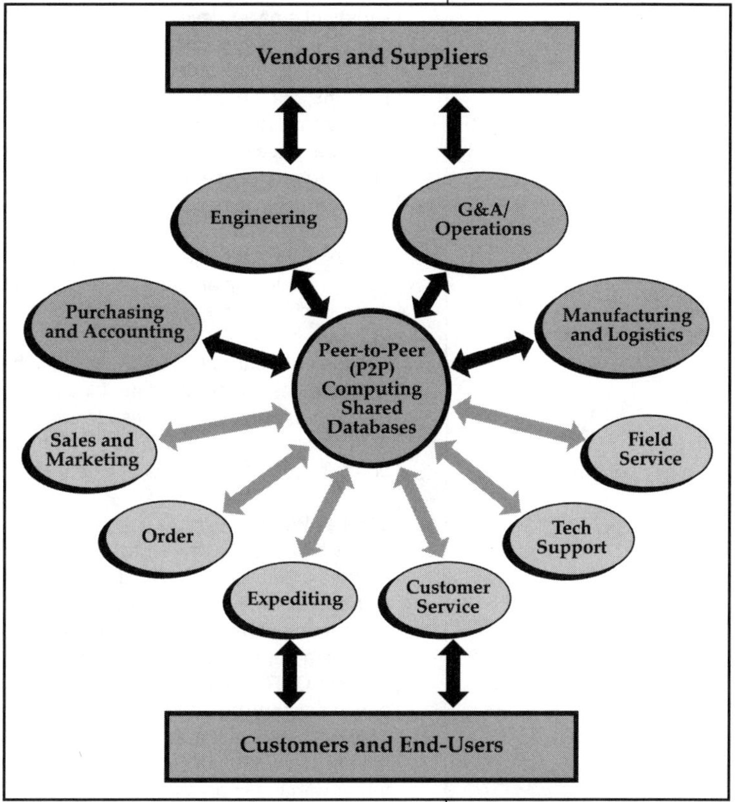

Figure 7-4. The architecture of the extended enterprise allows any unit of the company to perform its own portion of the customer service e-function.

The Tools of Online Customer Service

Call Centers

The modern call center is the heart of the customer service department. Though a call center does not, strictly speaking, have to be online, the Web can be used in combination with several other Internet-based tools to provide a high level of customer support.

In a modern, inbound call center, each person has a workstation in which a computer or terminal is connected to a network and a set of servers. The primary tools each phone operator uses are based on CRM (customer relationship management) concepts. CRM software provides the phone operator with a set of screens that make it possible to easily access all the customer or prospect information stored on the company's database servers. CRM software is designed to provide integrated access to all customer data in real time, including:

- Customer quality
- Order history
- Shipment history
- Service history
- Details of orders and problems
- Customer requirements data
- Customer status and credit information
- Notes and comments about the specific customer

(continued)

There are many vendors of CRM software, including megaplayer Siebel Systems (www.siebel.com). Siebel, PeopleSoft (www.peoplesoft.com) and a variety of smaller developers are also providing CRM functionality in the form of online applications delivered by ASPs (application service providers). In these ASP versions, the functionality remains the same as, or close to, the traditional systems, but the data resides on the ASP's server and the user accesses it over the Internet via a Web browser interface. Some CRM packages are being created that have customized, non-browser-based interfaces but nevertheless use the Internet for accessing and storing the data. Since browsers are graphical and therefore quite slow compared with text-based screens, these non-browser interfaces promise greater efficiency.

ASP-based CRM software systems provide a way for small to medium-size companies to implement a CRM system without incurring the large, upfront costs of standalone systems. One of the most prominent CRM ASPs is Sales force.com (www.salesforce.com), which like much traditional CRM software is aimed specifically at sales. However, when the Salesforce.com technology is implemented successfully the focus is more on providing inbound and outbound call centers with ready access to information than on providing field applications of the technology. In fact, one of the most successful applications of CRM doesn't even involve sales at all but after-sale customer service.

CRM at Work

When a call comes into a call center, the CRM software displays the telephone number of the caller. If the database associates that number with a particular customer, the customer's information immediately appears before the CSR or operator. Other CRM software systems work by requiring the operator to input the customer's zip code or customer number. Within seconds of answering the call, the CSR has the customer's information on screen, including order history, service history, and notes and comments from previous customer interactions. This enables the operator to pick up where the last interaction with the customer left off. The CSR can come up to speed almost immediately and provide the customer with the impression that he "knows" her and the issues she is facing.

Many CRM implementations rely on providing operators with high levels of training so they can be given significant latitude in making decisions on behalf of the company. This means that the CSR is trained and trusted—given all of the customer's data—and is capable of, and expected to, resolve most of the customer's problems without referring to a supervisor or other staff member. CRM software also provides call timers and other means for gauging the CSR's efficiency while helping customers. In short, CRM software and call centers have evolved to the point where it is possible to provide customers with a high level of service (though not all call centers are created equal!).

Online FAQs

One of the most significant tools available for online customer service is simply to post a searchable FAQ (frequently asked questions) database on the company's Web site. This provides both customer service reps and customers with "self-service" support at all hours. International customers can thus get answers to their questions when the company is closed or regular support channels are not available.

There are two keys to making FAQs an extremely useful part of the customer service experience. First, it is important to include a comprehensive listing of both pre- and postsales questions. Post as many of the issues—and their solutions—that customers bring up as possible. Second, it is important that the FAQ database be easily searchable and browsable. A manufacturer doesn't have to give away its darkest secrets, but it should provide as much information as it prudently can. The easier it is for customers and call center operators to find the answers to their problems online, the lower the cost to the company of providing customer support.

Medium and large enterprises should integrate online FAQs with CRM databases. This makes it possible to for the customer service rep to post questions to the FAQ database instantly, as she handles each new situation. This

(continued)

FAQ database often evolves into an online "knowledge base." Microsoft provides one of the most impressive Internet-based knowledge bases at www.microsoft.com.

Either a properly designed and carefully maintained knowledge base or a simple FAQ posting will reduce the number of calls the customer service department receives many times. For one company, one of the authors designed an online manual and FAQ system that permitted the company to grow over two hundred percent in sales while maintaining the same level of customer support with the same departmental head count.

"Live Help" and Other Chat Clients

One problem with call center-based customer service is that only one operator can speak to one customer at a time. In response, several companies developed Web-based "live help" technologies, which grew out of the "chat" technology used in bulletin board systems, Instant Messenger, and Internet Relay Chat. (A great demo of this technology can be found at www.livehelponline.com). Some of these chat clients will plug directly into a company's Web site, enabling a single customer service rep to easily handle simultaneous chat sessions with up to four customers. These chat sessions can even be outsourced offshore. Though a strong command of the English language is important for spoken customer service, oral English skills are not an issue when customer service is provided in written form.

Virtual Meetings

Several companies occupy niches in the "virtual meetings" space. Microsoft, for example, offers NetMeeting (www.microsoft.com/netmeeting), while Webex (www.Webex.com) and PlaceWare (www.placeware.com) also have substantial offerings. One of the most important applications of these virtual meeting clients is "virtual on-site customer service." It is now possible for a customer using a wireless notebook computer, digital camera, and Webex or PlaceWare to remotely show the customer service rep pictures and software (such as programming tools or diagnostics) at the customer's site. The customer can even easily show the CSR a video of a problem piece of equipment at its remote site—all without requiring a technician or engineer to jump on an airplane and solve the problem in the field. There are clear limitations to the technology, but the advent of P2P ("point-to-point" serverless networking) and WAP-enabled (wireless application protocol) devices will make it easier to do "virtual service calls" in the not-too-distant future.

Discussion Lists and Listservs

Since the early days of the Internet, many customers highly knowledgeable about the Web have asked for the ability to talk with their peers. For this reason, one of the most useful sources of customer service information has become online discussion lists and forums. The precommercial Internet had an entirely separate function, called USENET, devoted to these discussion groups. In those early days, many technology and software companies attempted to sponsor their own USENET newsgroups to provide customers and employees with a way to communicate service and support information interactively. USENET was fairly strict about allowing such commercially sponsored newsgroups, but the Internet continued to evolve.

Today, many organizations continue to set up their own discussion forums and e-mail mailing lists, for both sales and customer service. These forums and discussion lists are available to subscribers via e-mail and are often called "listservs," after the most commonly used software client for producing and maintaining them. Listserv®, marketed by L-Soft International (www.lsoft.com), is one of the most successful Internet software application ever created. Companies often outsource the hosting of their discussion lists, but they assign the monitoring of the groups as a specific responsibility of sales personnel, product managers, and customer service reps.

CHAPTER 8

e-Manufacturing: Tying the Internet to the Factory Floor

We've discussed how manufacturers can put the Internet to work in a variety of business functions, but what about production? What is "e-manufacturing?" In a nutshell, we are talking about the application of Internet technology to plant floor operations, from factory devices through MES and ERP systems (see Figure 8-1).

Accurate, real-time data is the heart of manufacturing. Sharing it throughout the organization quickly and efficiently is every manufacturer's goal. Providing customers, suppliers, and distributors with access to all the relevant bits would be a dream come true.

Which brings us back, once again, to the Internet. Deployed within an organization, it can provide plant floor personnel with real-time views of remote data. Embedded within factory equipment, it can produce smart sensors and intelligent machine tools. Connecting one company to another, the Internet can foster collaborative designs and enhance just-in-time deliveries. When combined with SCM and CRM systems, the manufacturing e-business makes it possible for employees and partners to drill down into the corporate intranet to check product availability, place orders, and track shipments, anytime and anywhere (see Figure 8-2). For manufacturers, the Internet may be the greatest thing since interchangeable parts!

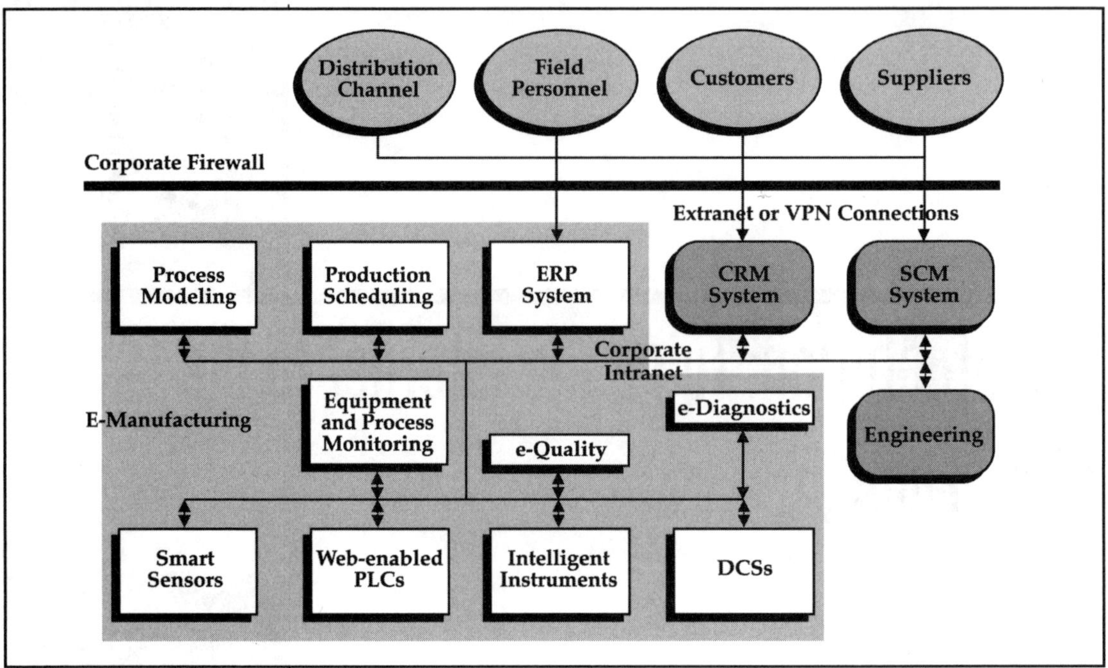

Figure 8-1. e-Manufacturing integrates Internet technologies throughout production, from plant floor devices to corporate business systems. An integrated e-business allows employees and partners to drill down into the corporate intranet to check product availability, place orders, and track shipments.

e-Real Time

Soon after Web browsers hit the street back in the mid-1990s, a small software developer in Massachusetts created a revolutionary new technology. The company discovered how to send real-time data across the Internet and display it in a Web browser! Initially, the company experimented with the technology in the online music business, but soon thereafter, Dr. Fred Putnam, the firm's president, applied his findings to the world of data acquisition and control through his firm LABTECH (www.labtech.com). Since then, sensing and displaying factory floor data over the Internet using standard Web browsers has become routine. Nearly all data collection products contain some form of this capability, from SCADA and HMI packages to SPC and instrument calibration programs (see the sidebar "Net-based Data Acquisition and Control," page 142). Now, instead of having to call from your office in Boston to your plant in San Jose, you can launch your favorite browser, type in a password, and download up-to-the-minute sensor data or SPC reports.

e-Manufacturing systems rely on the usual core technologies to gather and distribute factory data—namely, the external TCP/IP-based Internet, computer servers, and standard Web browsers. In addition, because most manufacturing takes place within a company's brick-and-mortar walls, e-manufacturing systems assume that factory equipment resides on Ethernet-based networks, which are increasingly tied to corporate intranets. The final piece of the puzzle, the thousands of specialized software packages that supervise the collection, analysis, and display of factory data, are nearly all browser compatible, to one degree or another.

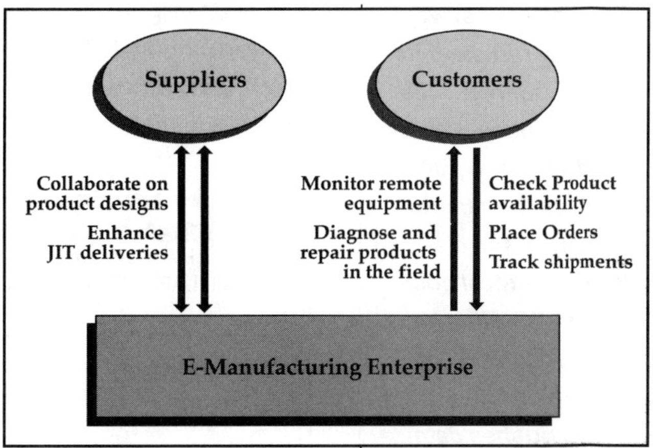

Figure 8-2. e-Manufacturing allows an enterprise to interact with customers and suppliers at a level beyond simple e-commerce, improving the efficiency of core engineering and manufacturing processes.

Collecting and Sharing Data

Most Web-based monitoring of equipment and processes is done using standard dial-up connections through the company intranet, via a modem and phone, or through ISDN lines. For security-conscious applications where password protection alone won't suffice, firms can connect to the Internet through a virtual private network (VPN). A VPN is a software-hardware combination that uses tunneling to encrypt the information being transferred so hackers can see only a cacophony of undecipherable characters. While no piece of equipment can provide one hundred percent guaranteed security (only physically disconnecting from the Internet will accomplish that!), the risks of using a VPN to send sensitive data online are more than outweighed by the cost advantages. VPNs are much cheaper than private networks, which use expensive, leased lines.

There are several ways to display remote data over the Internet. One approach uses a Web-based server to pull data from the remote instrument and display it on a local computer workstation by way of special client software. Another method pushes data directly from the instrument or PC-based host system, bypassing altogether the need for a separate Web server. In either case, a single browser page can display data from multiple devices, regardless of their location.

As an example, consider the process of sending automatic alarms from a Web-enabled machine on the factory floor directly to a pager or cell phone. With the "push" method, the machine or sensor simply sends e-mail to the address of the pager or phone as needed, for example, 8005551212@messages.myphone.net. A short text message

There are several ways to display remote data over the Internet. One approach uses a Web-based server to pull data from the remote instrument and display it on a local computer workstation by way of special client software. Another method pushes data directly from the instrument or PC-based host system, bypassing altogether the need for a separate Web server. In either case, a single browser page can display data from multiple devices, regardless of their location.

can easily be displayed on the phone or pager, such as "Machine 204-1. Warning. Thickness gauge reports paper thickness out of parameters."

Net-Based Data Acquisition and Control

by Frederick A. Putnam, Ph.D.,
President, LABTECH Corporation

Almost as soon as the Internet was launched, scientists asked, "Wouldn't it be great if one could access sensor readings remotely via the Internet?" And, in fact, this capability was first made available, albeit in a rudimentary form, as early as the 1970s! Since then the industry has made much progress. Today, Internet-based sensing is a mature technology that is readily available commercially.

Web Browsers

State-of-the-art methods for Internet-based sensing and process control involve viewing sensor data directly from Web browsers. Here, the viewer or "client" is most often a PC with an Internet connection. The sensor is wired to a "server" computer, which may be a PC, an instrument, or even a microprocessor integrated into the sensor (see Figure 8-3).

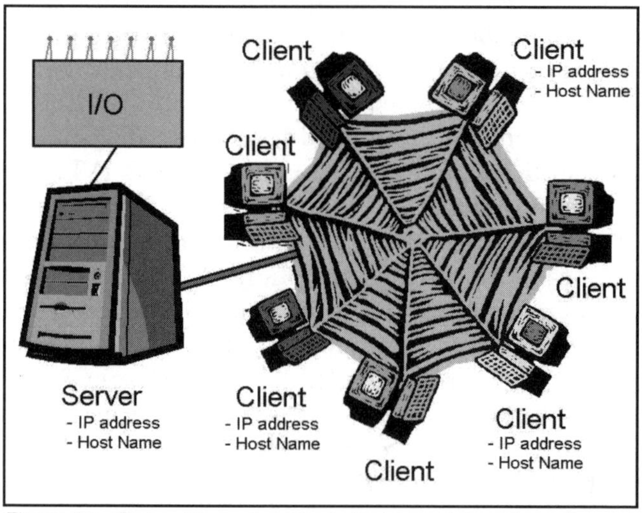

Figure 8-3. State-of-the-art methods for internet-based sensing.

The key piece of information that allows the server and client to send and receive information is their Internet addresses. The address is analogous to the phone numbers we use for telephone communication. Internet addresses can usually be expressed in two equivalent forms, the *host name* and the *TCP/IP address*. The easiest of these to use is the host name. This is an alphanumeric string, which is much easier to remember than the TCP/IP address. For example, the host name of our company's sensor data server is ltserver.labtech.com. The last part of this string, labtech.com, identifies the *domain name* of our company on the Internet. The TCP/IP address corresponding to this host name is 209.61.86.107. This series of numbers is the low-level form of an Internet address. On the Internet, a network of *domain name servers* performs the task of translating host and domain names into their TCP/IP addresses when they are needed.

(continued)

Web-based Servers

Hewlett Packard pioneered an interesting approach to Net-based data acquisition by embedding a small Web server in a microprocessor that is connected directly to a sensor. The Web server creates a Web page that contains sensor readings; the data can be collected periodically or each time there is new data. Whenever a client requests this Web page, it is sent out over the Internet to that client, who can see the data values. This elegant technique is now incorporated into many products.

Unfortunately, the Web server approach has limited flexibility and efficiency. This is because a Web page that contains the sensor data must be constructed entirely by the small Web server that is attached to the sensor. This server is an embedded system that has limited processing power and a small amount of read-only memory. As a result, the Web page layouts are difficult to change. Also, while the Web pages that are served are fine for viewing the data, they are not well suited for clients who wish to parse the data values and perform additional processing and/or local storage of the data. Finally, the Web server approach is inefficient because it requires that an entire Web page be sent every time the data needs to be updated. This means it is sending much more data than is necessary.

Application Servers

A more efficient and versatile approach to viewing direct sensor data is a system in which the serving of the sensor data and the serving of the Web page are kept separate. Here, there are two separate data streams. The sensor microprocessors send *only* the actual sensor data across the Internet. They do not serve any Web pages, so they are not Web servers but rather specialized application servers—sensor data servers.

The Web page is served separately from any corporate Web server, or even from the local hard disk of the client PC. This is beneficial because the Web pages will be served from the same server that all the other nonsensor corporate Web pages are served from.

This "separation" approach uses a Web page plug-in to receive and render the sensor data. (A plug-in is a small, simple software component that runs on the client PC when it is loaded by a Web page.) There are several different ways to implement plug-ins: Microsoft ActiveX controls, Netscape plug-ins, and Java applets. Any one of these can be used to retrieve the sensor data from the server and then render the data in the Web page as well as provide any other processing or storing of the data that might be necessary.

This application server architecture was pioneered by LABTECH in early 1996 and is now commercially available in several of the company's products.

The Future

Looking into the future, one can see great potential in wireless Internet-based data acquisition and control using handheld devices like the Palm Pilot (www.palm.com) and Handspring Visor (www.handspring.com). Also, expect to see actual closed-loop control over the Internet shortly. Such technology is here today, and research and pilot programs are underway. Security and reliability concerns are the primary reasons why it is not in wider use today. However, these issues will gradually be solved, as the Internet becomes more hardened and capable of more mission-critical work. The developers of these ultrareliable extensions to the Internet will constitute the new e-commerce industry.

Across the board, we are seeing the emergence of more intelligent factory-floor devices, from smart sensors to Web-enabled programmable logic controllers (PLCs). PLCs, in fact, are just now coming up to speed with their PC-based instrument counterparts. Both can display real-time trends and colorful graphics using standard Web browsers, and both can send e-mails when prompted by alarm conditions. These capabilities allow factory- floor devices to take on a more expanded role. Besides I/O, they can now keep track of historical data and monitor maintenance schedules. With the advent of personal digital assistants (PDAs) that can interface with corporate databases, wireless technology is even beginning to emerge on the factory floor. Several companies have introduced wireless documenting calibrators that support both PDA and Web interfaces.

An Issue of Quality

The Internet is best known as a resource that can pull together departmental information from a wide variety of locations. The area of quality control is no exception. In the last few years, the Web has become an important resource for collecting quality information from a multitude of corporate databases and file types.

Using a Web browser, you can now view and analyze operations from any location that has a networked interface—remote PCs, thin client terminals, handheld devices, even wireless phones. You can gather data from quality logs and generate reports using all the familiar analysis tools—X-bar, R, s, process capability, p, np, and more. Today's Web-based quality software can even provide access to other Web pages, online documentation, CAD drawings, flow charts, process diagrams, and other Web applications. As a result, anyone with the proper security access can instantly participate in improving plant floor processes. This makes it easier to identify and resolve problems, and speeds the delivery of information throughout the enterprise.

Almond producer Blue Diamond uses Northwest Analytical's (www.northwestanalytical.com) NWA Quality Analyst Web Server as part of their almond production and quality tracking system (see Figure 8-4). The system collects data using Wonderware's (www.wonderware.com) HMI/SCADA package and feeds it into a SQL database. Quality Analyst reads the data from the database and sends control charts and reports to Web pages that are developed and

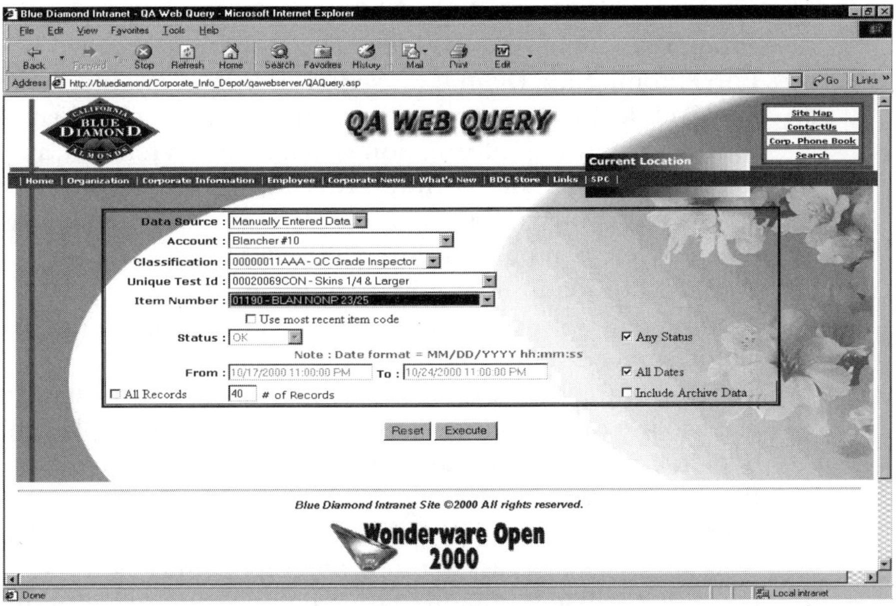

Figure 8-4a. Blue Diamond has put Northwest Analytical's Quality Analyst Web Server to work examining moisture levels in blanched almonds. The initial interface allows personnel to select from a variety of tests.

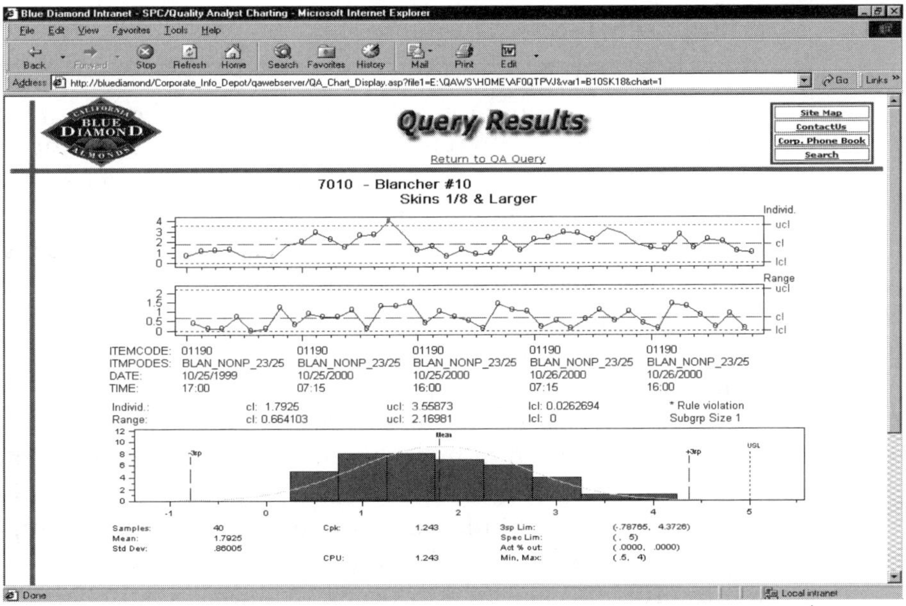

Figure 8-4b. Blue Diamond provides a variety of SPC charting functions online in real-time for up-to-date status on such product attributes as almond size and moisture.

maintained in house. A Visual Basic user interface for navigating the system was developed by an internal team led by Bob Hitomi, Blue Diamond's senior industrial engineer.

"Web technology is an important enabling technology for manufacturing information systems," notes Jeffrey Cawley, vice president of Northwest Analytical. "The Web provides a powerful extension of our plant networks and lets us extend the reach of monitoring and reporting with little extra cost or manpower. It is much less expensive and requires less overhead than adding extra seats to a standard client/server application."

One of the difficulties in making it easier for nonstatisticians to use operations data has been the need to display the data in easy-to-use formats. One company, VisiBit Corporation (www.visibit.com), has used the latest research in human pattern recognition to develop a way to present large volumes of data in easy-to-interpret graphic images. This makes it possible for maintenance, engineering, accounting, and management to use a Web browser to look at very large data-point constellations and develop interpretations without special training or an advanced statistics background.

e-Diagnostics

Firms can also use the Internet to monitor the health of remote equipment. Both from within an organization and from without, the Internet can be used to watch for changes in performance that might indicate that a repair is needed, even before the production line needs to be shut down. This provides suppliers with an incredible opportunity to get closer to their customers. A supplier's field service technicians can now monitor equipment data to anticipate customer needs and know which parts need to be brought to the plant and when. If customers need to interface with one of the technicians, they have their choice of a wide variety of communication channels—phone, fax, e-mail, or a live Web-based conferencing link. The bottom-line benefit is that suppliers can be more proactive, can anticipate and solve customer problems before they occur. In addition, suppliers can use the Internet to collect detailed historical performance data from networked equipment. This information can be fed back into the design engineering process, resulting in more accurate diagnostic tools and better products.

Varian (www.varian.com) has taken e-diagnostics one step further. The company can send schematics and drawings electronically to its customers by way of a video helmet. The documents are displayed in the helmet's eyepiece and can be immediately viewed by either customers or Varian field service personnel.

Several companies are experimenting with the idea of equipping field service and maintenance technicians with "wearable PCs" like those made by Xybernaut (www.xybernaut.com). Using wireless LAN and Web connections, these "wearable computers" are now small enough that staff can wear them while climbing tanks and ladders. This makes access from the field to the information in the corporate SCM and CRM systems not only practical, but feasible.

Before long, we'll also see smart sensors monitoring their own internal health. When their performance begins to go awry, they'll contact their operator (regardless of where he or she is) and order a replacement part from the factory. If it's a software upgrade, they'll just download it then and there. When the part ships from the factory, an order will be placed with the service department to schedule installation: minimal downtime, minimal customer hassle. This is already becoming a reality in the bulk chemical industry where storage tanks are monitoring their own inventory, ordering replacement chemicals, and tracking usage data. Nalco, the huge water treatment chemical vendor, believes so strongly in the future of this type of interconnectedness that it has taken a major equity position in Xsilogy Corp. (www.xsilogy.com), a small company that specializes in making low-powered, battery operated, Web-enabled sensors specifically for this market. Not only does Nalco want tanks that order for themselves, but also tank sensors that track the usage and health of the tank and sensor itself. What a great way for a vendor to build customer loyalty!

Nalco, the huge water treatment chemical vendor, believes so strongly in the future of this type of interconnectedness that it has taken a major equity position in the company Xsilogy Corp. (www.xsilogy.com), a small company that specializes in making low-powered, battery operated, Web-enabled sensors specifically for this market. Not only does Nalco want tanks that order for themselves, but also tank sensors that track the usage and health of the tank and sensor itself. What a great way for a vendor to build customer loyalty!

Web-based Control

While Web-based equipment and process monitoring is gaining acceptance throughout the industry, Net-based control is an altogether different story. In most cases, manufacturers' concerns about security and bandwidth have kept control tied to the old reliable workhorses of the plant floor, PLCs and Distributed Control Systems (DCS).

In R&D labs, however, we are starting to see increased use of the Internet for remote control of experiments. One of the earliest, in fact, was a small garden at the University of

Southern California. Anyone could connect to this "tele-garden" via the Web and water a plant, pull a weed, or apply fertilizer. Unfortunately, participants were more interested in using the technology than growing a healthy garden. The plants soon became overwatered and overfed, a situation we definitely do not want to encourage on the factory floor.

In like vein but with better results, Dick Morley (www.barn.org), the inventor of, among other things, the PLC, has been demonstrating his "Web-enabled tractor" to show that even real-time control of active manipulators is possible over the Internet.

New Development Tools

But while Web-based control is slow off the mark, Web tools are increasingly infiltrating control devices. Java, for instance, is a platform-independent programming language created by Sun Microsystems (www.sun.com) as a fast, simple way to program for the Web. Right from the start, the Web faced the challenge of providing a common platform that could integrate a mix of computer platforms and software. Sun developed Java to be a portable, architecture-independent programming language that could be universally deployed—program once, read everywhere. Because it has no pointer architecture, no direct access to memory, and includes built-in features for Web access and security, Java has come close to delivering on its promise. (The only problem has been the slow speed at which a Java-based application initially loads on a Web browser).

Because of this success, we're seeing Java put to work within control devices themselves to solve the same situation on the plant floor that is solving on the Internet—integrating a mix of equipment platforms and software applications that the typical manufacturer has running throughout its enterprise. Schneider Automation, for instance, is using Java with its Web-enabled PLCs. Invensys's Foxboro products group is using a series of Java applications to connect subsystems within its Intelligent Automation series of DCSs for SPC, maintenance, and advanced control functions.

The XML protocol is also making headway on the plant floor these days, albeit at a higher level than direct control of real-time I/O. Vendors such as Opto-22 are using XML to tag and transport data between databases or from databases to Web pages. The beauty of using XML is that you no longer need special interfaces or device drivers to transfer

I/O to a display. The OPC Foundation has even created a working group on XML to look at ways to better achieve interoperability in situations where performance is not the most critical factor.

Invensys' Wonderware group, for instance, is using XML in ActiveFactory, its data-mining software. XML provides a way to transfer data between Wonderware's SQL server-based databases, which are filled with plant floor information, and all the potential users in a company. Wonderware and most of the other HMI software developers have realized that they must do more than just collect and display real-time equipment data. Now customers want to integrate and analyze the information to gain a more comprehensive view of their organization's operations. It is demands like these that will drive the XML standardization process. Nevertheless, we still expect that it will be several years before the significant advances that are being made trickle down to small and medium-size manufacturing enterprises.

Now customers want to integrate and analyze the information to gain a more comprehensive view of their organization's operations. It is demands like these that will drive the XML standardization process.

ANSI/ISA-95.00.01-2000

In the meantime, the Standards Group of ISA (Instrumentation, Systems, and Automation Society, www.isa.org) is not standing idle. They are working on defining a new standard for integrating control systems within the enterprise. Titled ANSI/ISA-95.00.01-2000, the new body of work defines the interface between manufacturing control systems and business logistics. Ultimately, ANSI/ISA-95.00.01-2000 aims to provide enterprise software developers and equipment manufacturers with a core definition of all the relevant manufacturing interfaces. This will enable the enterprise to be integrated from the bottom to the top in the easiest, most cost effective way possible. Parts 1 and 2 of the standard, *ANSI/ISA-95.00.01-2000 - Enterprise-Control System Integration: Models and Terminology* is an official standard (see sidebar "ANSI/ISA-95.00.01-2000").

ANSI/ISA-95.00.01-2000

ANSI/ISA-95.00.01-2000 bases its functions on several existing manufacturing models: the Purdue Reference Model, the MESA International Functional Model, and the equipment hierarchy model from the IEC 61512-1 (ANSI/ISA-88.01-1995) standard. These describe the basic functionality of the manufacturing enterprise. From these models, the ANSI/ISA-95.00.01-2000 authors are slowly but surely defining the interface between the factory production system, as defined by these models, and the business activities of the enterprise (R&D, finance, marketing-sales, and corporate management), as shown in Figure 8-5.

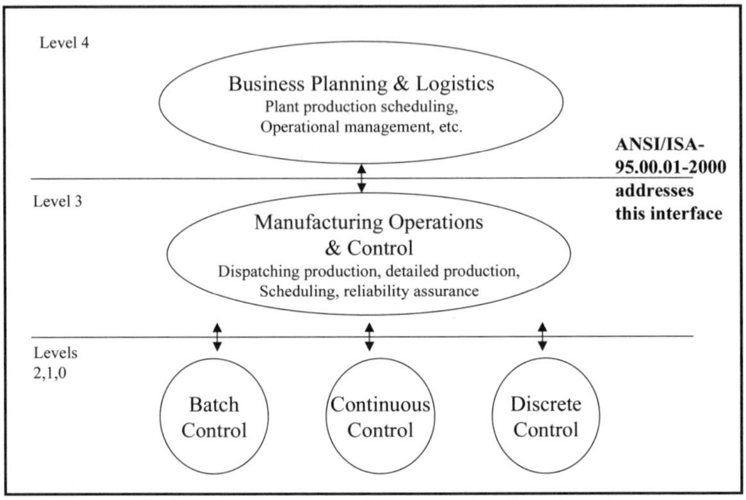

Figure 8-5. ANSI/ISA-95.00.01-2000 defines the interface between manufacturing operations and control, and business planning and logistics. © 2001, ISA—The Instrumentation, Systems, and Automation Society

Part 1 of the standard provides a comprehensive definition of resource types (personnel, equipment, material, process segments) and exchange elements (production capability product, production schedules, and production performance). For example, the definition of a material lot is an object that uniquely identifies a specific amount of material, either countable or weighable. A material lot may include the following:

- A unique identification of the lot
- The amount of material (count or weight)
- The unit of measure of the material (e.g., parts, kg, tons)
- A location for the material
- Any committed status of the lot

Part 2 of ANSI/ISA-95.00.01-2000 contains further examples to help clarify the definitions in Part 1. The example used throughout Part 2 involves the manufacturing of "widgets." The manufacturing process is described in terms of the resource types and exchange elements just mentioned. At the end of the standard, brief examples are given for electronic board assembly, food processing, crude oil transport, and more.

According to the editor of ANSI/ISA-95.00.01-2000, Dennis Brandl, director of enterprise initiatives for Sequencia, Parts 1 and 2 of the standard don't try to define a formal protocol for information exchange between systems. Rather, they provide a common base for these exchanges. They also offer a way to document the information exchanged, even if the exchange is done manually.

The committee decided not to formalize an XML description of the information, primarily because Web standards organizations are still battling over XML definitions. The World Batch Forum (www.wbf.org), a nonprofit professional organization, has taken the lead in developing XML schemas based on the ANSI/ISA-95.00.01-2000 standard. The schemas are made freely available to end users, system integrators, consultants, and vendors.

Design Collaboration

Leaving the knotty issue of standards behind for a moment, let's consider the unique environment in which today's manufacturers operate. Their customers are demanding instantaneous delivery and sport a "build-for-me" mentality. The challenge is to balance these customer requirements with the company's own need for happy employees and a profitable bottom line. To successfully compete in this quick time-to-market, build-to-order world, companies must focus on what they do best and fine-tune the rest of their operations to support JIT principles. As a result, it's not surprising to find that many manufacturers are now functioning more as stage managers than actors in that they coordinate what has become a multibillion dollar outsourced manufacturing market (witness the growth of the contract manufacturing giants Solectron and Flextronics).

It's not surprising to find that many manufacturers are now functioning more as stage managers than actors in that they coordinate what has become a multibillion dollar outsourced manufacturing market.

I first ran into this scenario back in the early 1980s, when I began doing some work with Apple Computer. On my initial tour of their state-of-the-art facility in Fremont, California, I was surprised to find that they manufactured almost none of the bits and pieces that go into Macintosh computers. I learned that Apple was one of the early pioneers in the computer assembly business. They ordered processors from Arizona and imported storage peripherals from the Pacific Rim. Once all the pieces were gathered, they creatively assembled them, in either California or Ireland, into one of the most leading-edge technologies of the day.

Focus on Brilliance

What Apple had learned at an early age is that the real brilliance behind its company was not in manufacturing a slew of computer peripherals, but in designing one of the most visually appealing computing technologies of the times. To this day, Apple continues to win awards for the innovative design work behind its CPU boxes and monitors.

The biggest challenge for such a distributed environment is coordinating the collaboration in a timely manner—among internal design engineers, between internal and external designers, between internal designers and external manufacturing engineers, and so on. The goal is to ease collaboration between in-house and independent design, manufacturing, and procurement personnel so manufacturers can—as i2 Technologies (www.i2.com) so aptly characterizes it—"Design Anywhere, Build Anywhere."

In the past, such efforts involved lots of travel, group meetings, faxed drawings, and long product development cycles. Today, we can no longer afford such luxuries of time and overhead. With product lifecycles diminishing to six months or less in some technology industries, we must learn to quickly deliver stand-out products, almost before our customer asks for them. And so we once again return to the Internet and the myriad tools for using the Web.

Sharing Product Data

The core requirements for a collaborative product development environment are a centralized database for storing all product data and a related set of logic that defines the rules for sharing data. From there, it's an issue of how people get to the information and to each other.

To develop corporate intranets, firms early on used collaboration tools that relied on groupware (see Figure 8-6). Dominated by Lotus Notes (www.ibm.com), such software provides employees with shared access to proprietary databases of information. Over time, developers applied Windows-based technologies to the task. This allowed corporations to use the more readily available Microsoft tool set and Windows-based browsers, which offered increased flexibility.

Today, standard browser-based front ends are providing employees with access to internal databases and applications, such as in-house directories, expense-reporting forms, project updates, and company classifieds. Such arrangements provide all the members of an organization with quick access to knowledge that used to be difficult to find. While the ROI on such "group memory" systems is hard to quantify, corporations are increasingly deploying them throughout their operations.

Conferencing

Online collaboration tools have also been around for several years, including Internet or VPN-based virtual whiteboards and Web conferencing services. As bandwidth becomes more widely available and organizations upgrade

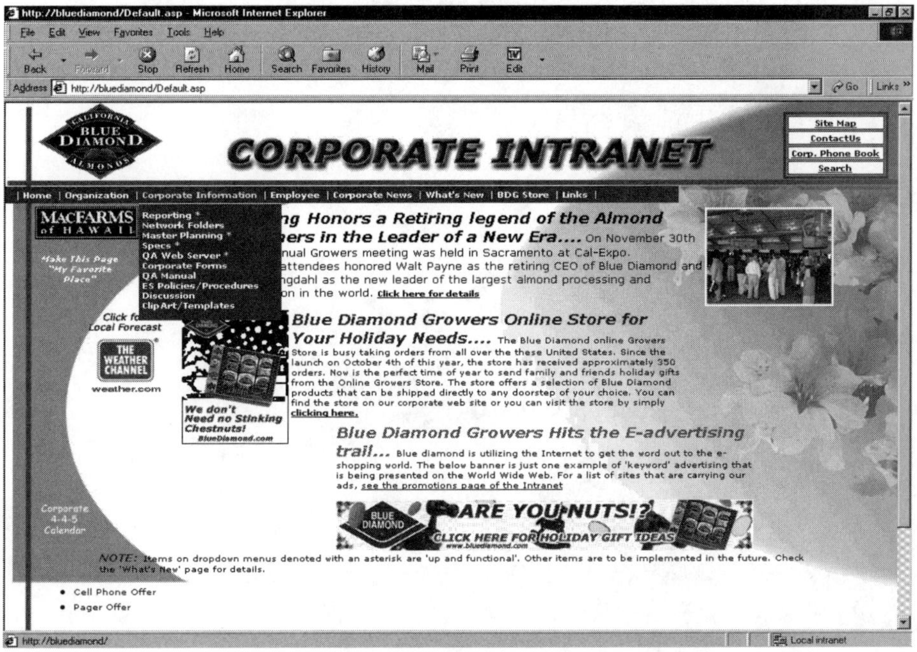

Figure 8-6. Blue Diamond provides access to selected pieces of the manufacturing process via the corporate intranet.

to T1 and T3 connections, client/server tools are making possible online conferencing environments that integrate voice, video, and data. Other approaches to conferencing make creative use of such traditional technologies as POTS (plain old telephone system) and ISDN (Integrated Services Digital Network). For instance, Motion Media Technology (www.motion-media.com) markets videophones with external ports that connect to laptop computers. Making conferencing calls no longer requires companies to endure the complicated setup or price of a large, room-size system. Instead, engineers can talk from their desks and instantly share documents and update drawings with ease using devices such as Motion Media's mm225 videophone (see Figure 8-7).

Even developers of plant design software have taken to the Web. Browser-based front ends allow engineers to view piping and instrumentation layouts. Schematics can be converted to Web-viewable formats, and engineers can query underlying databases to prepare reports and equipment spec sheets. To market its software, Intergraph (www.intergraph.com) provides the example of an instrumentation maintenance engineer who views a control loop and clicks on components associated with it. Doing so enables him to view linked equipment specs, maintenance histories, a 3D model, and instrument connection details.

Figure 8-7. Internet-based conferencing tools reduce development cycles by facilitating collaboration and document sharing. (Courtesy Motion Media Technology, Inc.)

Other Web-based tools enable design engineers, manufacturing engineers, and purchasing agents to easily collaborate earlier in the design process than was previously possible. Using these tools, they can create bills of materials and approved vendor lists and share supplier and team contact information. One of the simplest of these tools is eQuill (www.eQuill.com), which gives users a way to mark up a Web page exactly as they would a Word document—by using change tracking and sticky notes and by drawing on the page (circles, arrows, figures and the like).

Integration with Business Functions

Given the incessant push for higher quality, faster time-to-market, and lower costs, it was inevitable that today's increasingly complex manufacturing businesses would be forced to undertake large-scale integration efforts. What better way to improve operations than to have a clear view of internal business data?

This push began in the 1970s as large companies adopted materials requirements planning (MRP) software, which makes it possible to automate inventory management and control. In the 1980s, these companies turned to MRP II, manufacturing resource planning, which consisted of software for integrating inventory control with order processing and shipping. In the 1980s, businesses moved to automate all their in-house "back end" functions with enterprise resource planning (ERP) software, which combined MRP II functions with finance, human resources, and plant maintenance.

While the tightly integrated enterprise information architecture ERP promises is highly desirable, reality paints a slightly different picture. Largely the domain of Fortune 1000 companies, ERP systems are complex, unwieldy database monoliths with complicated user interfaces and complex code structures. They typically cost tens of millions of dollars and years to complete. Only now, nearly twenty years since their introduction, are they beginning to be sized and priced for medium-size enterprises. Thus small manufacturers are forced to continue working with MRP II systems on the back end and PC-based sales management databases, like ACT, on the front end.

The biggest problem ERP faces, however, is the perception that it is only relevant for integrating "back-end" functions. With the advent of the Internet, the push has been on the "front end"—developing improved sales, marketing, and customer service systems that better integrate the enterprise with the customer via the ubiquitous Internet. Experts say ERP vendors have changed and have been working hard in the past few years to expand their offerings and adapt to the new world of collaborative commerce. Still, they only provide a little more than half the CRM, SCM, and e-procurement functionality of vertical application developers. The result is an information gap that will cost corporations billions of dollars to fill.

Experts say ERP vendors have changed and have been working hard in the past few years to expand their offerings and adapt to the new world of collaborative commerce. Still, they only provide a little more than half the CRM, SCM, and e-procurement functionality of vertical application developers.

A few companies are stampeding to the forefront. PeopleSoft, for instance, devoted the bulk of its resources over two years—$500 million and two thousand developers—to redesign its enterprise software. Their new software is called PeopleSoft 8, which they describe as a "pure Internet platform for the collaborative enterprise."

Integrating the enterprise with the Internet is not the only issue. Most ERP vendors have focused their efforts on the discrete manufacturing industries where production revolves around a distinct product that has easily definable inputs and outputs. Process industries, on the other hand, must struggle with complex inputs, batch operations, multiple units of measure, different formulas, and more. A few ERP vendors, such as Oracle (www.oracle.com), claim to be targeting their collaborative commerce applications at both discrete and process operations. The jury's still out on their effectiveness.

Most ERP vendors have focused their efforts on the discrete manufacturing industries where production revolves around a distinct product that has easily definable inputs and outputs. Process industries, on the other hand, must struggle with complex inputs, batch operations, multiple units of measure, different formulas, and more.

Build to Order

The Internet is forcing manufacturers to face a new customer-driven world where building to order is no longer an option. Unfortunately, traditional manufacturing tools were not designed for a world where cost-cutting must be balanced with customer-centric planning and support. Traditional vendors are rapidly adding Internet functionality to their equipment and processes. But the big hurdle remains upgrading the manufacturing organization's internal information architecture. The goal is to make it possible for the disparate pieces of the enterprise to be integrated so employees, customers, and suppliers have a unified view of the entire business, from sales to customer service to shipping and receiving.

CASE HISTORY

Failure Analysis Goes e-Quality

Many of today's large electronics manufacturers provide some level of component testing services to their internal engineering and plant floor operations. These internal labs analyze failed components, help designers choose reliable parts, and work with manufacturers to optimize yields. Today, as companies push to build commercial businesses out of such services, they are looking for ways to differentiate themselves.

One major manufacturer, who prefers to remain anonymous at this early stage of the launch, has put the power of its internal infrastructure and the Internet to work for them. By cleverly combining an internal laboratory information management system (LIMS) and secure, Web-based access to customer projects, the organization created an innovative approach to helping manufacturers quickly solve product problems.

LIMS

The heart of the new e-quality system is a Windows-based LIMS. An Ethernet-based network connects together all of the major instruments in the laboratory, from scanning electron microscopes to high-end ESCA systems. Some procedures, such as pop-top analysis (i.e., pop the top off a chip and visually inspect it) require that data be entered manually. For the major instruments, data from the part under test is gathered, analyzed at the local computer workstation by an engineer or technician, and then sent to a central database. The central database organizes the information by customer and part. As the part analysis proceeds, a project report is developed that describes the tests performed and all the root causes of the problems that were found.

In the past, customers would have to wait until the entire testing project was complete before getting any results. They could always talk with one of the engineers to gain interim results, but nothing was documented until the final report was issued. This tended to put increased pressure on the lab's engineers as their customers anxiously awaited the test results. In some cases, manufacturing lines were down until the problem was found.

Today, the laboratory has put a new, customer-friendly face on its services by adding a Web-based front end to its internal systems. Customers can now go to the Web and log onto the new Internet portal of services. Through special extranet connections, customers now have 24x7 access to their product analyses. Interim data is posted online, along with images captured by the various pieces of test equipment. Engineers enter preliminary results into the system as soon as they are available. Customers can then call or e-mail the lab to discuss the results further. A final report is then posted online along with a hard copy available as a .PDF download.

It's too early to point to any solid financial results generated by this new business, although initial indications are good. The lab has turned its Web site into a true portal of information regarding failure analysis and product quality. Before a manufacturer becomes a customer, it can read dozens of detailed papers about the latest techniques for analyzing plastic packaging or determining stress failures in wave-soldered surface-mount components.

CHAPTER 9

The Technology of e-Commerce

As we discussed in previous chapters, industry analysts may disagree about the exact numbers, but all are quick to affirm that the future of B2B e-commerce is large, growing at a healthy clip, and highly competitive. In just a few short years, e-commerce has developed into a multibillion-dollar market with large IT budgets at work behind the scenes. According to the analyst firm Datamonitor, dotcoms alone will be spending over $9 billion per year on IT by 2006, the bulk of which will be on IT maintenance[1]. And you can add to that the even larger budgets of brik-and-mortar companies who are developing multichannel distribution systems that involve both the physical and virtual worlds. The end result is an impressively large opportunity for Web technology vendors.

So where is all this money going? We can divide the technology of e-commerce into two basic areas: the "store" and the "back office." The store is your Web site, your interface with the outside world. This is where you present, at a minimum, your product information, marketing brochures, sales updates, contact information, and methods for interfacing with customer service. Regardless of whether you take orders online, the Web provides your company with a remote, uniform interface to virtually any business function.

The back office includes everything that's going on behind the scenes of your Web site. It's all about internal operations, transaction processing, and connecting and integrating the Web site with your enterprise.

1. Demir Barlas, "US e-Commerce Booms in 2001," Line56, www.line56.com, July 11, 2001.

The Store

As we discussed in Chapter 4, on customer relationship management, you should start by making sure you've got the basics covered. Your product information and company contact information should be the first things placed online, and they should be in an easy-to-find, easy-to-read format. Beyond that, merchandising and selling skills are as critical to the success of your virtual store as they are to any brick-and-mortar operation.

What do we mean by "merchandising" in this context? We mean the way the site looks and feels, how it operates, how easy it is to navigate, and how well you present a compelling message to site visitors. Take an honest look at your current Web site. What is its purpose? How hard is it for customers to get the product information they need from your site? Do you want to take orders online? Are you trying to off-load customer service? Do you want to help distributors gain access to the latest inventory updates? It is important that the content on the site fit the goals and directions of your company. Ask yourself whether you're really delivering any "value add" to the customer.

Some excellent books have been written on designing compelling sites, such as the seminal work by Jakob Nielsen, *Designing Web Usability*.[2] The best sites involve a team approach in which product managers, graphic designers, and Web programmers working side by side with department managers. As with any technology project, you should bring in outside help if you don't have an established team with Web experience already in place.

The key is to remember that customers are still customers, whether they're interacting with your company online or off. Dirt, clutter, poor presentation, and ho-hum front-window displays cause customers to walk past traditional stores. Minimal content, amateurish design, lack of organization, and annoying clutter cause Web visitors to click by your site until they find what they need (probably at one of your competitors' sites). Here are some very important features to include:

1. Provide a site search engine and make it easy to find. Nothing will chase a customer away faster than the inability to find quickly the information he or she needs.

2. Jakob Nielsen, *Designing Web Usability*. Indianapolis, IN: New Riders Publishing, 2000.

2. Provide clear navigation guides at the top or side of the screen. Put them in the same place on every page. Consistency is important because you want the customer to become "comfortable" inside your store. This is the same reason department stores use the same iconography and merchandising design throughout all their departments.

3. Use Macromedia Flash animation, streaming audio, and pop-ups selectively as they usually take a long time to load and will hamper the customer's search for information. If you decide to use these technologies, always offer a lower tech alternative. For instance, PDF files are a convenient way to download and store data sheets for future reference. But they do not provide instant access. Make sure all product information is also available in standard HTML format. For business-oriented sites, focus on a clean, spare design with simple navigation and easy-to-use tools.

4. Make sure your site is accessible to people with disabilities. You never know whether a purchasing agent or major customer is disabled. Run your site through the Bobby engine at www.cast.org to make sure it is accessible to the disabled. Bobby will spot most of the common design errors, as well, not just ones that make sites difficult for disabled people.

5. Use site registration as a way to generate leads, but keep it optional. Never restrict access to product information. Use registration as a way to deliver "premium content". Testmart.com found that when they eliminated registration for everything except actual purchase, site usage went up over one hundred percent and bookings increased dramatically. To encourage visitors to register, offer a white paper or technology tutorial in exchange for some basic information from them.

6. Provide tons of content, after all, that's why the customer is visiting your site in the first place. And don't just focus on proprietary information. The more content the better. Establish your expertise in an industry by providing lots of technical content as well as links to related information. It will set you apart and boost your site's reputation as an industry portal.

7. Provide interactive tools, such as "Live Help," and make sure you provide a plethora of ways to contact

technical support and customer service. Remember that "computer-mediated" purchasing feels less personal than face-to-face interactions or even telephone sales. Little touches like personalizing your Web site can increase response rates by as much as fifty percent.

8. Adhere to the "three clicks" rule. Site visitors should be able to get to most everything they need within three clicks. A wide and shallow site is better than a narrow, deep one. In most cases, your customers are in a hurry and need to be able to find things quickly. You can confirm this by observing your own shopping patterns online.
9. Make sure your shopping cart works intuitively. Test and retest it on customers and partners.
10. Establish and adhere to a strict privacy policy. This is critical if you do business in the European Community where strict rules govern online privacy and the use of customers' personal information.

Push versus Pull

Several years ago, there was considerable debate swired around the concept of "pushing" content to potential customers. Many firms were intent on developing a technology that would allow companies to deliver information to potential customers even if they hadn't requested it. Initially, this involved integrating ads into Web browsers. As it turns out, however, the vast majority of Internet users do not want anything "pushed" to them online, even if it is information they might be interested in.

The most effective approach to delivering information online focuses on "pulling" visitors to the corporation's Web site. Once they are there, the site should deliver a full complement of product data, technology tutorials, and the like. This pull methodology relies on traditional techniques that have been adapted to the online world, such as e-mail newsletters and Web site advertising.

As an example, most companies send a bi-weekly or monthly e-mail newsletter to customers, partners, and prospects who have opted to receive it. This is also known as "opt-in." If a person has not registered with your site or the site of the agency from which you're renting names, you're likely to end up with a lot of angry people blasting you with e-mail about abusing their time with unwanted communications). The most effective newsletters focus on delivering technical content by using short lead-ins to information on the company Web site. The best part is that pull interaction is inexpensive, interactive, and highly successful.

Marketing departments were among the first to adapt pull techniques. As a result, online B2B marketing has jumped to $1.34 billion in 2000 and is growing at a 45.7 percent rate annually. And for good reason. When an e-news mailing was recently sent to the in-house prospects list of a US-based developer of open control systems it generated a forty-two percent response rate! The typical direct mail piece generates a one to two percent response.

Other proven methods for driving traffic to your Web site include the following:

1. Mentioning the site's URL in all company literature, on business cards, and throughout catalogs,
2. Advertising in the e-news mailings of related industry publications and associations,
3. Optimizing your Web site so it gains top rankings in the search engines, and
4. Highlighting Web content in traditional direct mail, such as four-color postcards.

HTML

The tools of a brick-and-mortar storefront include display cases, lighting, signage, and traffic pattern generators like end-of-aisle displays and the layout of the store itself. In an online store, the tools are virtual and made of software. They include markup languages for displaying text, programming languages for interactive features, and specialized hardware and software systems for security.

The basic tool of the Internet store, indeed of the entire World Wide Web, is hypertext markup language (HTML), as we have seen. HTML grew out of the graphical markup languages invented in the 1960s and 1970s during the creation of photocomposition devices for newspapers and book printing. Dr. Tim Berners-Lee, then at CERN in Switzerland, realized that by combining hypertext markup language with a unique Universal Resource Locator, or URL, he had a way of organizing and finding every discrete piece of information in the world. Thus was the World Wide Web invented.

Basically, HTML takes text or images and embeds controls and commands in it so that the text or image appears more or less as the creator intended in a "Web browser." HTML can be written in a simple text editor, like Windows Notepad, because it is not, itself a graphic language. It creates the graphics through the markups placed in the text.

For example, here is a simple markup followed by its result:

```
<B>Shari Worthington</B>
```

Shari Worthington

In this case, the <B> character says, "begin boldface" while the </B> character says "end boldface."

What is significant to remember is that each thought in an HTML document can be referenced via hypertext to another thought by using a link. For example, the code

```
<a href=http://www.waltboyes.com> Walt Boyes's Web site </a>
```

appears on a Web browser as follows:

Walt Boyes's Web site

This hypertext is linked to the address of Walt Boyes's Web site elsewhere on the World Wide Web.

By itself, HTML was a terrific invention. The creation of Web browsers, a standard interface that could read HTML, created the world of e-commerce as we know it. Unfortunately, HTML is "static," that is, it doesn't change dynamically. Such a tool is adequate for small Web sites of only a dozen or so pages, but when you are dealing with large quantities of information you need something more.

At its largest, the original Web site of ISA—The Instrumentation, Systems, and Automation Society (www.isa.org) contained over fifty thousand static pages of HTML! With this much information, it was impossible to keep the site current. Changes in one page that affected other pages had to be handled manually. This created a need for more powerful programming tools and ways to integrate databases behind the scenes.

Integrating Databases

The Common Gateway Interface or CGI was designed to output dynamic information from a database in a way that could be displayed on an HTML page. A CGI script is an executable program that resides in a special directory on the

Web server. The beauty of CGI is that CGI programs can be written in most programming languages, including C, C++, PERL, Fortran, TCL, UNIX and Linux shells, Applescript, and Visual Basic. All a programmer has to do is to provide a compiled version of the program and locate it in the proper directory on the server (usually called cgi-bin). Unfortunately, because CGIs are executable programs, serious security problems can occur. However, many of the firewall programs now available ameliorate these problems.

Larry Wall of NetLabs developed the first practical query language for use on the World Wide Web. Called PERL, for Practical Extraction and Report Language, it can locate and retrieve text strings in any HTML document. The current version, PERL5, is an open source document, like the Linux source code, and is available as freeware. In its early days, PERL served as the basis for many search engines. Today, it is used for much more than Wall originally intended. As one PERL expert characterized it, "It derives from the ubiquitous C programming language and its process, file, and text manipulation facilities make it particularly well-suited for tasks involving quick prototyping, system utilities, software tools, system management tasks, database access, graphical programming, networking, and World Wide Web programming. These strengths make it especially popular with system administrators and CGI script authors, but mathematicians, geneticists, journalists, and even managers also use PERL." [3]

To make even easier and more transportable programming tools, Sun Microsystems developed two programming languages, Java and JavaScript. Designed to be used with any operating system, Java and Javascript allow you to run the same code on a variety of OS's, including SCO UNIX, XENIX, Linux, and Windows NT. The utility of this tool for the Web was obvious and immediate. Without it, developers would face the significant challenge of creating separate programs for the multitude of computer configurations connected to the Net. As a result, many Web sites rely on some sort of Java or JavaScript programming for more advanced features. In the following JavaScript example, a series of messages scroll on the information bar in the lower left hand corner of a Web browser.

3. Jarkko Hietniemi, Frequently Asked Questions. Comprehensive Perl Arhchive Network, www.perl.com/CPAN-local/doc/manual/html/pod/perlfaq1.html.

```
<SCRIPT LANGUAGE="JavaScript">
<!-- Beginning of JavaScript Applet -------------------
function scrollit(seed)
{
    var m1 = "Welcome to MarketingPractice Consultants...";
        var m2  = "Consulting, Authoring, and Seminars...";
    var m3  = " Next Update: April 15th... ";
    var m4  = "Real World Issues...Real World Answers!.... ";
    var m5  = "Contact us now!"
    var msg=m1+m2+m3+m4+m5;
    var out = " ";
    var c   = 1;
    if (seed > 100) {
        seed--;
        var cmd="scrollit(" + seed + ")";
        timerTwo=window.setTimeout(cmd,100);
    }
    else if (seed <= 100 && seed > 0) {
        for (c=0 ; c < seed ; c++) {
            out+=" ";
        }
        out+=msg;
        seed--;
        var cmd="scrollit(" + seed + ")";
          window.status=out;
        timerTwo=window.setTimeout(cmd,100);
    }
    else if (seed <= 0) {
        if (-seed < msg.length) {
            out+=msg.substring(-seed,msg.length);
            seed--;
            var cmd="scrollit(" + seed + ")";
            window.status=out;
            timerTwo=window.setTimeout(cmd,100);
        }
        else {
            window.status=" ";
            timerTwo=window.setTimeout("scrollit(100)",75);
        }
    }
}
// -- End of JavaScript code -------------- -->
</SCRIPT>
```

As you can see, JavaScript is a "non-programmer's" programming language. It runs on any Web browser, on any operating system, without modification.

The ongoing disputes between Microsoft and Sun led Microsoft to develop another component-based programming language, similar in many respects to Java, called ActiveX. ActiveX is an increasingly popular programming language among industrial automation software vendors and is seeing some application online.

One of the newest kids on the block in the Web languages department has really taken the industry by storm. XML (for eXtensible Markup Language) is a way to "tag" hypertext items with a standard description so companies can use them to exchange information between companies. XML now serves as a core technology in all electronic marketplaces and is even being used by many corporations in electronic data interchange (EDI) applications. It is also proving to be an invaluable tool for integrating a company's Web site with the back office.

In May 2001 the W3C (the World Wide Web Consortium, the "governing body" of the Web) published a complete XML Schema as a recommendation (W3C calls its standards "recommendations"). Already, Apache, IBM, Microsoft, Tibco, and several other firms have produced XML Schema implementations. Other XML tools are within a year of becoming standards themselves, such as XML Query. When XML Query is a completed standard, it will bring the power of SQL to XML files. It will then be possible to build queries around collections of XML files that may contain unstructured data like documents or HTML Web pages.

Security

For the first few years of e-commerce, the bugaboo of e-commerce was the fear that somehow the customers' information would be stolen. As we discussed in Chapter 3, this fear is both justified and manageable. One of the earliest ways to manage it was to provide some means for securing information transfer over the World Wide Web. e-Mail security relies on PGP (Pretty Good Privacy), which was introduced in the early 1990s by Phillip Zimmerman. Based on the public key encryption method, PGP uses two encryption keys, one a freely disseminated public key and the other a private key that is used to decrypt the messages. This type of encryption is so difficult to break in practice that the U.S. government sued Zimmerman to try to keep PGP off the market.

In the mid-1990s, Netscape (now part of AOL Time Warner, but then the overwhelming leader in Web browser technology) used a similar public key encryption technique to produce SSL (Secure Sockets Layer) technology for its

Netscape Navigator Web browser. Microsoft soon incorporated SSL into its Internet Explorer browser. SSL and its companion, S-HTML, provide a securely encrypted connection between a client and a server over which data can be transmitted. Data is encrypted at the origination point, then decrypted at the receiving server. Web page URLs that provide a Secure Sockets Layer connection start with "https:" instead of "http:"

The introduction of SSL and S-HTML made it possible to transmit credit cards, purchase orders, and other sensitive information over the Internet without special connections or private networks. SSL and S-HTML are official standards approved by the Internet Engineering Task Force (IETF). In the late 1990s, the emergence of 128-bit encryption made security even tighter. Today, it is much easier to steal credit card records from dumpsters behind shopping centers than to steal them as they move across the Web.

From Text to Multimedia

The original Web browser, Lynx, was a text-only technology that was unable to handle graphics or pictures. Then, in the mid-1990s, Dr. Fred Putnam created the concept of streaming audio. He and other entrepreneurs, such as RealNetworks and later Microsoft (with Windows MediaPlayer), made it possible to receive streaming audio and video over the Web. Assuming you have a sufficiently broad Internet connection, today's Web browsers can display complete multimedia presentations. Napster took things a step further by making it possible to distribute multimedia files to millions of different computers, all of which act as virtual servers and effectively increase available bandwidth.

Animating the Web

Bandwidth became a much sought-after commodity as advertisers, marketers, and broadcast and mainstream media industries pushed for ways to make the Web browser more like a television. Their goal was to find a way to "push" programming to computer users, as does TV and cable. Early forays have focused on delivering short animations via Web sites.

The most significant tools in this arena are Macromedia's Flash and Shockwave. Flash has become the defacto standard for animation programming on Web sites today. An object-oriented animation language, Flash is so easy to use that even children can put it through its paces. Advanced graphic artists use it to provide stunning animations direct to the browser. Not surprisingly, the challenge is that Flash requires a relatively high-speed connection to the Internet, at least a 56kbps dial-up. As a result, it is not always recommended for B2B commerce, since many users have lower bandwidth connections and are usually not interested in such entertainment. In B2B, Flash is appropriate for producing your training videos or product demonstrations. But you should keep them short so the download times don't scare your customers away. Macromedia's other animation toolkit, Shockwave, allows a Web page to display multimedia objects, such as a video or audio file, through a standard browser window.

Also finding increasing use in B2B commerce is VRML (Virtual Reality Markup Language). Amarkup language similar to HTML, VRML produces *hyperspace* the way HTML produces *hypertext.* This gives developers the ability to display objects in three dimensions in a standard browser window. In the context of e-commerce, this means that three-dimensional objects can be displayed on a two-dimensional screen, allowing the customer to "walk" around the product or maneuver through the virtual plant. Early uses for VRML have included virtual tours of real estate and manufacturing plants, as well as virtual trade shows where the attendees can "walk" into a booth and look at products as if they were actually walking down the aisle of a trade show.

Human Factors

Just as with a brick-and-mortar store, the designer of a World Wide Web storefront must give human factors special attention. Professor Donna Hoffman and Dr. Tom Novak of Vanderbilt University's Own School of Management have written extensively about computer-mediated marketing.[4] They have expended a tremendous effort codifying what works in Web marketing.

4. D. L. Hoffman and T. P. Novak (1996), "Marketing in Hypermedia Computer-Mediated Environments: Conceptual Foundations" *Journal of Marketing,* 60 (July), 50-68 and other, more recent works.

According to them, regardless of your site's purpose, it is extremely important that you design it so that users are able to browse in comfort and with ease. As in the survey by the National Association of Manufacturers, who found that a mere one percent of manufacturers' Web sites can be considered world-class. Similarly, most of the sites we investigated for this book were appallingly engineered from a human factors perspective. Web sites are more than a collection of product literature. They need to be designed with the site's visitors in mind. It's amazing how many obvious issues have been overlooked by most manufacturers. For instance, typefaces need to be large enough and clear enough to be easily read. Sites with dark backgrounds, while visually arresting, are difficult to read and navigate. The bottom line is that it is not enough to merely know how to use the tools of the Web. To produce a successful commercial site you must also understand the principles of design and how they relate to the computer-mediated environment.

The Back Office

As in a brick-and-mortar environment, inventory, billing, fulfillment, and customer service and support are critical to the success of a Web endeavor. In fact, such "back office" functions are even more important for an online store because the Web shields customers from the inner workings of the company. When you are face to face with a customer in a traditional physical store, you can address difficult situations as they arise. You can even give customers a peek behind the curtain into the back of the store, if necessary. When your store is online, however, your customers are often in the dark. The more access you can give them to the information they need, the more effective your Internet presence will be.

We discussed clients and servers in Chapter 2 (for more information, you can find an excellent Web-based tutorial on Web servers at Webcompare.internet.com/Webbasics/index.html). So, we'll spend a moment instead on transaction processing, that is, the methods for securely transmitting information from one point on the Internet to another. When implemented correctly, this is a crucial set of technologies that enables manufacturers to provide customers with an increasingly intimate view into selected internal operations.

After the construction of the Web site itself and the underlying server, the next most important e-commerce technology is the transaction-processing system. There are many types of transaction processing systems, all of which focus on providing safe and secure transmission of electronic data. Such systems tend to be divided into manual, semi-automatic, and automatic systems. Many of the most recent developments in transaction processing are occuring in transactional middleware. These are software systems that process transactions on behalf of a client and route the information through many diversified systems, such as from an NT server to a UNIX server to a mainframe.

For small sites, the most common transaction-processing system is a simple e-mail-based "order advisement." A customer enters the required order information, including credit card numbers and other payment details. The system combines this with the product information in the shopping cart and inserts it into an encrypted e-mail that is sent from the Web server to the site owner's designated e-mail address. The order information in the e-mail is then manually entered into the merchant's order-processing system. Once entered, an order acknowledgement is manually e-mailed or faxed to the customer. While this is typically only done in small sites, it offers an easy way for manufacturers to test the waters of e-commerce. In fact, Best Buy's Web site worked *exactly* this way during the 2000 Christmas season. We advise you to try it only with fairly low-volume products, though.

Fully automatic systems, like those used by most e-marketplaces, capture the order data entered by the customer, check stock, enter the order, process the payment option selected, and provide a nearly instantaneous order acknowledgement to the customer. Behind the scenes, such systems interface either directly or indirectly with the inventory control and accounting system of the manufacturer or distributor. Daily pricing and inventory updates are automatically uploaded to the Web server. While this approach is straight-forward to implement (note we didn't say inexpensive), it can frustrate customers and partners because it is inevitably out of date. Direct, fully automatic interfaces to the inventory control system enable the Web server to process orders instantaneously and produce real-time inventory numbers. Unfortunately, at the moment, such systems are very expensive to implement.

Whatever shape it takes, a direct interface between the Web storefront and the extended enterprise enables a company to fully realize the savings and added value of e-commerce and e-business. As technology continues to evolve, more options are becoming available for the small and medium enterprise. As shown in the McDaniel Controls case study in this chapter shows, you do not have to be a large company to put e-commerce to work for you.

Payment Systems

by George Matyjewicz

Paul Volcker, former chairman of the U.S. Federal Reserve Board, has said that "A global economy requires a global currency." Nowhere is this truer than in e-business. While B2C Internet commerce has settled on several "credit card-type" payment systems as well as on credit cards themselves, the requirements of business are demanding better and more global solutions.

The United States has five percent of the world's population and probably ninety-five percent of the world's credit card users! Devised in the 1960s, credit cards have outlived their usefulness, especially on the Internet. They are ill equipped to make payments over a public network and are the source of common fraud at levels not yet fully revealed. In addition, many companies do not permit their purchasing functions to use credit cards for anything but the most minor of maintenance expenses. The demographic distribution of credit cards also militates against their becoming a global purchasing tool. For example, in the Muslim community it is forbidden to use credit cards to purchase goods if there is another alternative available, preferably gold. Then there are the "unbankable" people who have neither credit cards nor checking accounts.

Other payment systems became popular because of specific events and needs. Manufacturers selling to importers in a foreign country needed a way to transact business and get paid properly. Thus, letters of credit (LCs) were born (in the late Middle Ages). An irrevocable letter of credit transfers funds when the importer is assured that the goods are manufactured to specification; the manufacturer is guaranteed payment when the goods reach a designated port. Letters of credit have not changed dramatically for several hundred years. The Muslim world uses a similar process called guaranteed insurance contracts (GIC), which basically say, "I will purchase your products for $100 million, and I guarantee I have the money, and an insurance company will guarantee that I have those assets to cover the purchase."

EFT and EDI

Technology and the Internet introduced new methods of payment. Electronic funds transfer (EFT) allows a person or company to electronically transfer money from one location to another worldwide. When it was first introduced decades ago it was revolutionary—a company could transfer funds to any destination in one to seven days!

Similarly, electronic data interchange (EDI) made it possible for companies to do business electronically when they had agreed beforehand to credit terms. Purchase orders, requisitions, and lot releases could all be transmitted electronically, if each end of the transaction was enabled for EDI.

Then came the Internet, where everything is immediate. Suddenly EFT was not so speedy and it was costly. EDI was (at least temporarily) declared "dead." The Internet has hatched cash substitutes from smart cards to electronic payment companies. At first, there was a gaggle of startups interested in generating electronic currencies; many of these failed, primarily because shoppers do not want to use a special currency or an account just to buy a book online. Smart cards are said to be the way of the future, yet at least one major company gave up on them after only one year.

(continued)

Electronic Currencies

The latest technological advance is digital currencies. PayPal,* which has grown to nine million subscribers in thirty-six countries, allows two parties to transact business online. The buyer uses PayPal to purchase from the seller and the funds are supposed to be immediate and nonrepudiable. PayPal has proved to the world that an online payment service could gain a sufficient mass of customers. Unfortunately, its current business model appears to have some major flaws. PayPal is based on weak account-based technology such as credit cards and the automated clearinghouse (ACH) system. Many PayPal merchants and shoppers have stories to tell about the drawbacks of the PayPal system. For example, several merchants and individuals have had their accounts frozen because of suspected fraudulent use of a credit card that was not even part of their transactions. For businesses concerned with payment and cash flow, this is a major concern. Nevertheless, PayPal has shown that electronic currencies are indeed possible and can be successful.

e-Business requires simple, easy, and secure cash and credit transfer. As businesses globalize, and as the Internet forces pricing terms and delivery to do the same, doing business in multiple currencies becomes bothersome. The European Community's euro currency was designed to deal with this, but since it is backed by the varying strengths of several national currencies, it is not doing as well as expected.

One of the most interesting trends to watch in the next few years will be the divorce of currency from individual governments. For example, a relatively new company named Standard Reserve Holdings Limited has developed an extremely interesting take on a global payment system.** Through its subsidiaries and network of agents, it already has account holders in one hundred fourteen countries worldwide. Standard Reserve has a one-for-one, gold-backed currency with debit cards, which allows customers to extract funds at over four hundred fifty thousand ATMs worldwide. Its solution appears to work well for the Muslim community, since the currency is gold and no interest is charged. Thus, global businesses now have a way of providing immediate, non-reputable payments, and multinational companies can handle employee payroll or suppliers' payments without the delays of wire transfers.

* www.paypal.com
** www.standardreserve.com. The Web site for its subsidiearies can be found at www.standardtransactions.com

George Matyjewicz, managing partner of GAP Enterprises, Ltd. (www.gapent.com) has been writing a column for Gifts & Decorative Accessories magazine since 1996. He is also moderator of e-Tailer's Digest, a moderated discussion list that reaches retailers in forty-nine countries worldwide. His email address is georgem@gapent.com.

CASE HISTORY CASE HISTORY CASE HISTORY

McDaniel Controls, Inc.

McDaniel Controls (www.mcdanielcontrols.com) is a small provider of specialty pressure gauges, primarily to the petrochemical industry. Located in Luling, Louisiana, McDaniel Controls has made a significant attempt to integrate back-office functions into its corporate Web site. In this case study, we'll walk through the major elements of its pricing and inventory status system.

In Figure 9-1, we see the McDaniel Controls home page. It provides immediate access to a wide range of interactive user features. Note the "Gauge Search/Inventory Lookup" feature.

As Figure 9-2 shows, McDaniel provides two methods for querying the inventory database. One uses a step-by-step process that guides you through a non-trivial number of product options. The company also provides a search by part number function for distributors and experienced customers who know which part they're looking for.

Figure 9-3 shows the company's Gauge Search interface, which is the primary means of interfacing to their inventory system. Note that it is a simple interface that, as Figures 9-4 through 9-11 show, eventually produces pricing information and inventory status. If you change a product parameter, the inventory quantity changes, as do the available gauge types. For security reasons, the inventory is only mirrored on the site. The date and time of the most recent inventory upload is listed so customers and distributors can make appropriate decisions about ordering items that show low inventory levels.

A password-protected portion of the McDaniel site is reserved for customers and distributors. Here they can securely access information about their orders. The company also provides engineering support, including CAD drawings and specifications that are customized to the specific needs of the individual user. All in all, McDaniel's is a simple yet functional site with a clean user interface that gets customers the information they need.

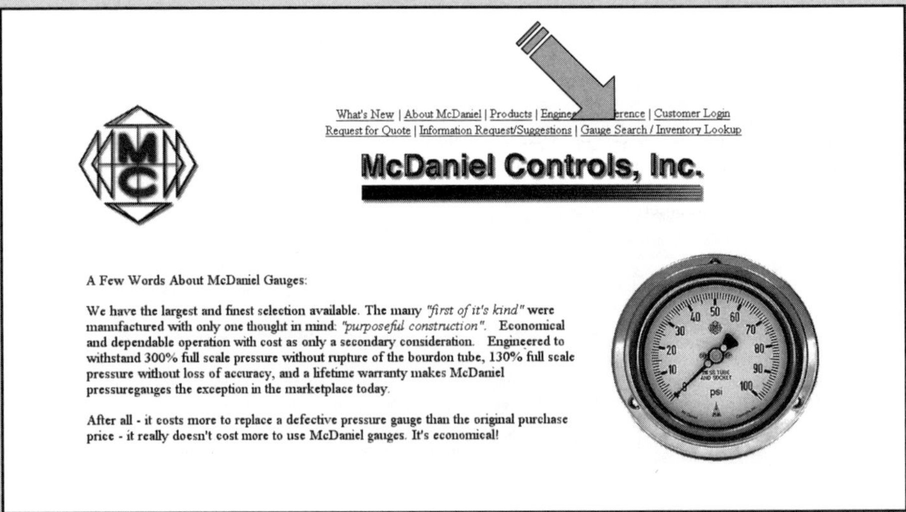

Figure 9-1. McDaniel Controls, Inc. gauge search/inventory lookup feature

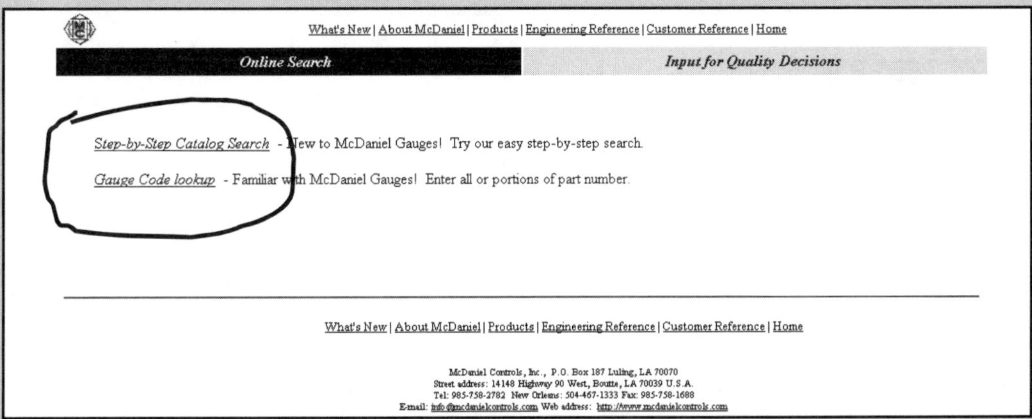

Figure 9-2. McDaniel Controls, Inc. two method query for the inventory database

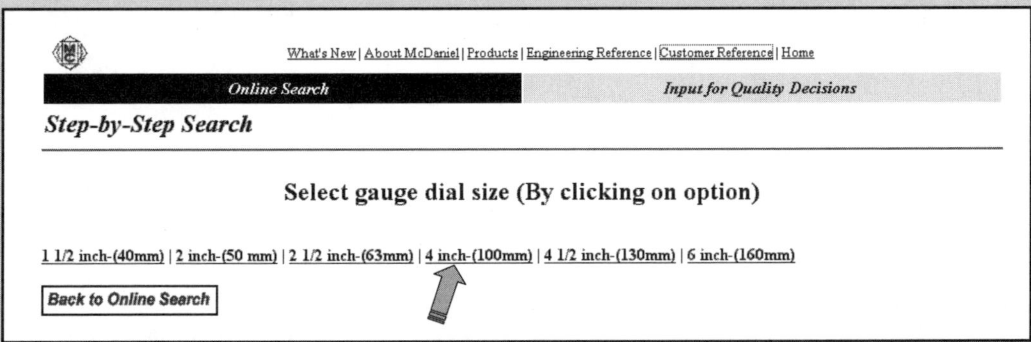

Figure 9-3. McDaniel Controls, Inc. gauge search interface

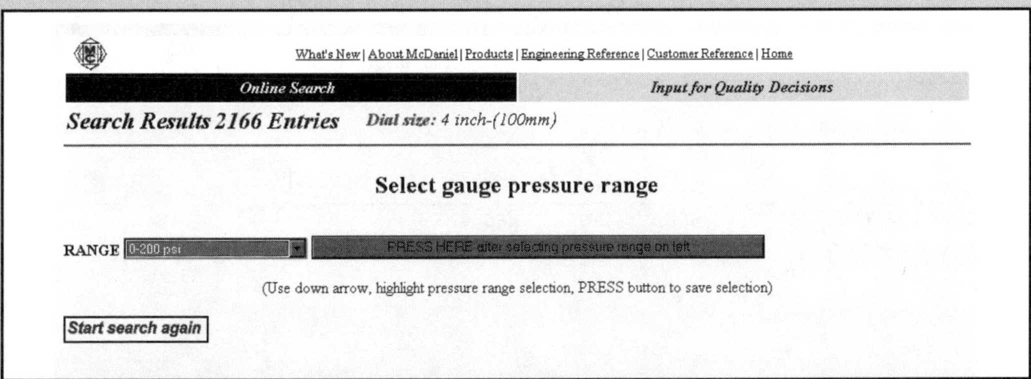

Figure 9-4. McDaniel Controls, Inc. gauge search interface step 1

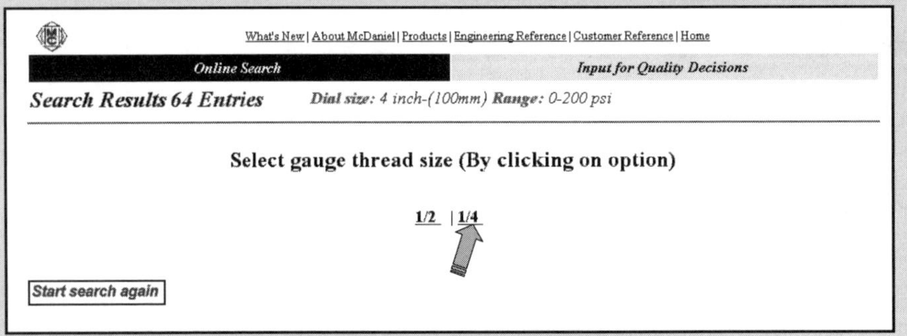

Figure 9-5. McDaniel Controls, Inc. gauge search interface step 2

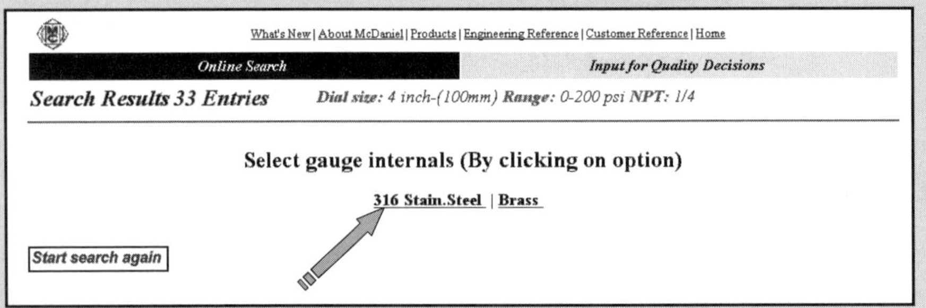

Figure 9-6. McDaniel Controls, Inc. gauge search interface step 3

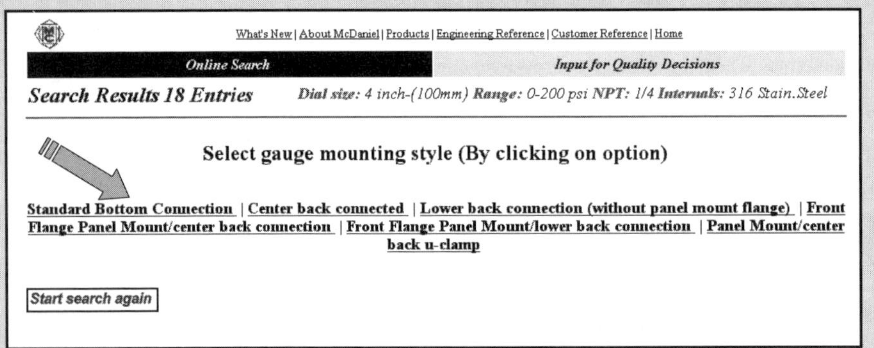

Figure 9-7. McDaniel Controls, Inc. gauge search interface step 4

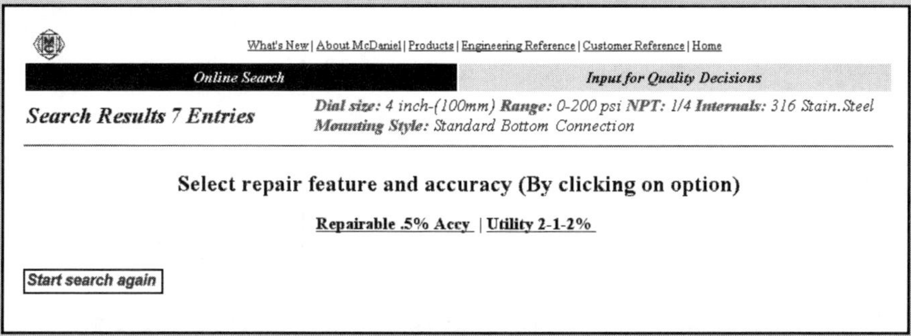

Figure 9-8. McDaniel Controls, Inc. gauge search interface step 5

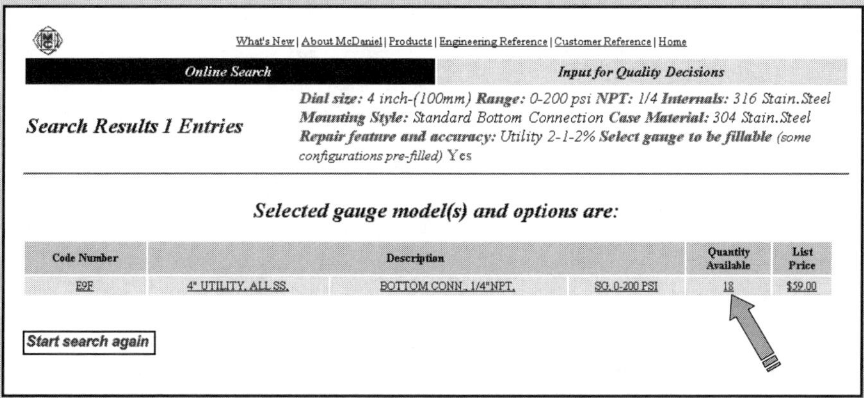

Figure 9-9. McDaniel Controls, Inc. gauge search interface step 6

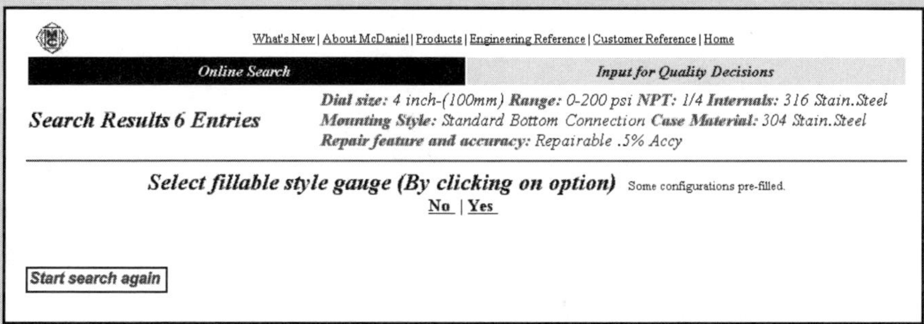

Figure 9-10. McDaniel Controls, Inc. gauge search interface step 7

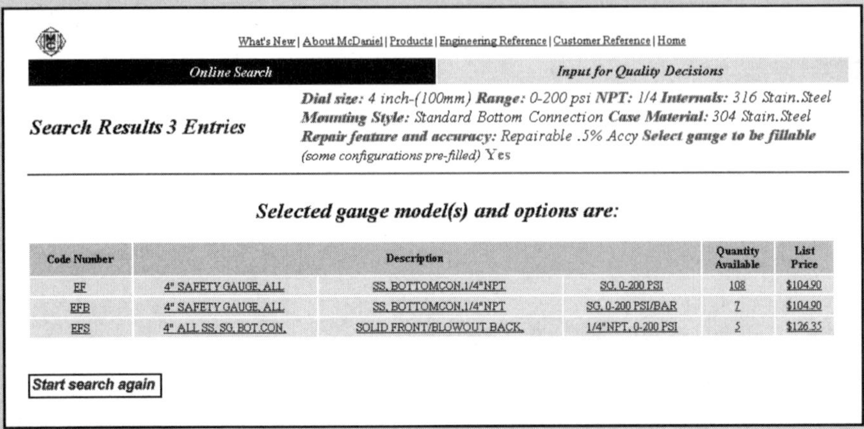

What's New | About McDaniel | Products | Engineering Reference | Customer Reference | Home

Online Search — *Input for Quality Decisions*

Search Results 3 Entries

Dial size: *4 inch-(100mm)* ***Range:*** *0-200 psi* ***NPT:*** *1/4* ***Internals:*** *316 Stain.Steel* ***Mounting Style:*** *Standard Bottom Connection* ***Case Material:*** *304 Stain.Steel* ***Repair feature and accuracy:*** *Repairable .5% Accy* ***Select gauge to be fillable*** *(some configurations pre-filled)* Yes

Selected gauge model(s) and options are:

Code Number	Description			Quantity Available	List Price
EF	4" SAFETY GAUGE, ALL	SS, BOTTOMCON,1/4"NPT	SG, 0-200 PSI	108	$104.90
EFB	4" SAFETY GAUGE, ALL	SS, BOTTOMCON,1/4"NPT	SG, 0-200 PSI/BAR	7	$104.90
EFS	4" ALL SS, SG, BOT.CON,	SOLID FRONT/BLOWOUT BACK,	1/4"NPT, 0-200 PSI	5	$126.35

Start search again

Figure 9-11. McDaniel Controls, Inc. gauge search interface step 8

CHAPTER 10

The Future of e-Business in the Industrial Enterprise

Arthur C. Clarke, the well-known futurist and science fiction author beloved by countless engineers, has always loved technology. His invention of the principles of satellite communication from geostationary orbit won him many awards and, along with his dozens of sci-fi adventures, demonstrate his passion for advanced technology. Along the way, Clarke insightfully noted that people tend to overestimate the short-run impact of technological change while simultaneously underestimating the long-run impact. It's a dangerous trap with serious consequences.

Take the case of fiber-optic networks. On June 18, 2001, the *Wall Street Journal* reported that about thirty-nine million miles of fiber optic cable stretch beneath the U.S. in a variety of obscure and not so obscure locations. Lying beside highways, railroads, and natural gas pipelines, there's enough cable out there to circle the earth 1,566 times. About $90 billion worth of that fiber has been installed in just the past four years alone. But the astonishing part is that a mere three percent of that fiber is actually in use. The rest lies dormant, representing what telecommunications guru George Gilder calls "dark fiber." As a result, the telecommunications industry lies battered and bloody. Service providers and chip manufacturers alike have suddenly seen the telecommunications revolution turn sour.

What happened to bring telecom growth to a crawl? As Arthur C. Clarke could have predicted, it was a case of the short-run impact of the Internet being vastly exaggerated. It had all started innocently enough...

The Growth of the Web

After the introduction of the World Wide Web in 1994, the Internet began to grow at a phenomenal pace. According to industry experts, usage doubled every two to three months, a pace never before experienced even in the most ubiquitous of technologies—radio and television. With the availability of easy-to-use tools such as Web browsers and plug-and-play modems, nontechies were finally able to take advantage of the power of the Internet and move beyond the obscure, complex world of ARPANET, FTP, and Telnet.

All seemed right with this new path. The commercial sector took the Internet by storm. Web sites popped up faster than organizations could gather good content. After a few years, companies realized the importance of the Internet and began to develop good content. e-Mail became the number-one form of business communication in high-tech organizations; other industries followed their lead. Most industry pundits were amused to find themselves forced to revise the Internet's growth statistics every month or so. Before long, over two hundred million people were using the Internet on a fairly regular basis. Things were moving fast.

But in all the excitement something was missed. Sometime during 1999, four short years after the introduction of the Web, the number of new users signing on to the Internet slowed. In fact, in comparison to the heady days of the mid to late 1990s, new growth nearly came to a stop. But no one was watching. Instead of usage doubling every few months, the Internet's growth slowed to a more respectable twenty to thirty percent per year. It was inevitable.

But recent evidence indicates that the Internet rarely ever grew at the incredible rates claimed. Even during the dotcom heyday it appears that online traffic was only doubling every twelve to fourteen months. A recent issue of the *Industry Standard* underscored the impact of the misleading data: "It's impossible to know just how many leaky business plans were launched on this premise."[1]

1. Jason Krause, "*Has the Net Stopped Growing*?" Industry Standard, July 2, 2001, p.32.

The Telecom Tailspin

As we now know, several major industries, backed by the deep pockets of overly optimistic venture capitalists, didn't anticipate the slowdown. The dotcoms and telecommunications companies were hit first and hardest. In late 2000, the industry pundits suddenly realized that there weren't enough people buying goods online (human psychology, including human buying habits, is slow to change). By early 2001, there weren't enough bodies to absorb the excess capacity of the high-speed networks around us (though the last mile to our doors is still mostly copper wire!). Without continuing growth in the number of people online, the dotcoms and telecom providers couldn't cover the cost of their extremely large up-front investments.

As the stock market reacted, so too did the suppliers to the telecom industry and those suppliers' suppliers. Network owners retrenched, some by adding standard voice communications to their portfolios to round out revenue. But not enough found profitable opportunities to offset their fiber losses. As a result, manufacturers of fiber cable and network equipment took a major hit. Once they slid, so did their suppliers—mainly, chip manufacturers and contract manufacturing firms. The reverberations are still being felt today.

So what does all this mean for the industrial enterprise? Should you stop investing in the Internet? Should you retool back to the old methods of communicating with customers, partners, and suppliers? Nothing could be further from the truth.

Information Must Be Portable

Information runs the world of business. To be useful, that information must be mobile. The Internet is the best method we have of turning our static and hierarchical organizations into the flexible entities they need to be to profitably face the future. The answer lies in moderation. In developing a reasonable business model for integrating the Internet into the manufacturing organization. Incrementally investing in new technologies while keeping an eye on costs and profits.

e-Marketplaces have diverted many an innocent manufacturer's attention in recent years. Hundreds of millions of dollars have been spent on automating external processes between buyers and sellers while internal functions floundered and continued running on phone and paper and faxes. This constituted a major disconnect with the highly automated processes of e-markets and was especially ironic for the automation and controls industry. The livelihood of equipment vendors depends on automating the processes of their customers' manufacturing operations, yet these same vendors have been extremely slow to automate their own internal functions. In fact, in 2000, fewer than twenty-five percent of the potential vendor partners for TestMart.com's sensors e-store had more than a rudimentary electronic data interchange capability. Not surprisingly, the e-store didn't launch.

But it wasn't just an issue of the millions of dollars in sunk costs. The e-marketplace hullabaloo derailed corporations in their work on the most effective business tool of the Internet era—the corporate Web site. As we mentioned in Chapter 1, less than one percent of all manufacturers can claim to have developed a world-class Web site, yet the Web has become the preferred method of interaction for many customers. This is the age of self-service. Your customers are as pressed for time as you are. And they insist on being provided with the information they need at the time that best suits their schedules. In many cases, that means your customers are perusing your Web site long after the last salesman has gone home. In fact, studies have shown that about thirty percent of a person's time online at home is devoted to work-related activities.

What's coming next in the world of e-business? What new technologies will soon have an impact on your organization and the way you conduct your company's affairs online? To answer those questions, let's take a closer look at the technologies of the future: telecommunications, the Web itself, software agents, and biometrics.

Telecosm

For all the problems besetting the telecom business and its overbuilt network of fiber-optic cables, there is a lot to be said for the new incarnation of this century-old industry. George Gilder fondly refers to the next wave of telecommunications as the "telecosm," the combination of fiber optics, high-spectrum communications, and wireless networks that will soon take over the world.

Proponents of the telecosm see a completely new infrastructure for telecommunications just around the corner. In his seminal work *Telecosm*, Gilder recounts a conversation he had with the inventor of the single-mode optical fiber, Will Hicks. They animatedly discuss a world where bandwidth is unlimited. Instead of copper wires and switches, computer terminals and satellites, they envision a world where fiber-optic and wireless networks carry millions of times the bandwidth of existing wires. Every phone can handle real-time images and therefore every conversation can become a videoconference.

According to Gilder, fiber optics and wavelength division multiplexing (the process of dividing light into finely tuned wavelengths that can carry gigabits of data—parallel processing, if you will) will make hay of the traditional seven-layer network architecture. Everything will be eliminated but the physical layer for transmitting bits through a fast but dumb network. Packets of data will bounce through a connectionless medium of fiber-optic strands or through air (for cellular connections). Like Ethernet, the intelligence of the network will be in the terminals at either end, where packets are decoded and interconnected.

Peering into the future of wireless, Gilder expounds, "The most common personal computer is already a digital cellular phone, a teleputer. There are three hundred million cellphones in use, increasingly digital ones, and just two hundred million personal computers. Soon your cellphone will be more powerful than your conventional PC. It will be as portable as your watch and as personal as your wallet. It will recognize speech and convert it to text. It will plug into a slot in your car and navigate streets. It will collect your news and your mail and if you wish, it will read them to you. It will browse the World Wide Web and download information as needed. It will tell you the weather at your destination. It will conduct transactions and load credit into an encrypted credit meter on its microprocessor or onto a chip on a smart card, which can be used like cash. It can pay your taxes or help you legally avoid them. It will take pictures and project them onto a wall or screen or dispatch them to any other teleputer. It may not do Windows. But it will do doors—unlock your car door, open your front door, unroll your garage door."[2]

What it all comes down to is the inevitable push for more useful and more advanced technologies. Metcalfe's Law, developed by Robert Metcalfe, inventor of Ethernet, states that the value of a network rises in proportion to the power of all the devices attached to it. What this means is that the

2. George Gilder, *Telecosm: How Infinite Bandwidth Will Revolutionize Our World.* New York: Free Press, 2000, p. 83.

more computers, handhelds, and plant machinery are connected to a common network, the more freely information will flow both within and between enterprises. The more information there is available to employees, even to other equipment, the higher will be the quality of the decisions made. The better the decisions, the higher the profits and the greater the long-term viability of the corporation.

The Future Web

One of the biggest frustrations of today's Internet is that the software and servers behind the scenes do not really understand what we mean as we hyperlink our way through the billions of Web pages and links now available. When we query a search engine for information on "PLC," for instance, many of the links will reference "plc" as a type of British corporation rather than the beloved programmable logic controller.

For the answer to this problem, we turn to the inventor of the World Wide Web, Tim Berners-Lee. According to Berners-Lee, the Web has evolved into a world of documents for people rather than a universe of information that can be automated for use by computers. This makes it hard for us to do much more than read the information available online. In the near future, he believes the Web will be structured differently. Software agents will roam from page and page, exchanging information in a more meaningful way. Berners-Lee calls this new form of the Web the "Semantic Web."

If today's Web can be likened to a gigantic book based on HTML, the Semantic Web will look more like a giant database built around two core technologies: XML and RDF (resource description framework; see Figure 10-1). XML creates the language of tags that provide common labels for Web pages. RDF is a way to describe the meaning of terms in a triplet form that computers can understand—along the lines of a sentence's subject, object, and verb. As described in the May 2001 issue of *Scientific American*, RDF assigns properties to various items, including people or Web pages. These properties can be assigned values, such as an association to another person or another Web page.[3]

As an example, consider a customer perusing your Semantic Web site. Your data acquisition equipment can be described in terms of a variety of applications, each of which

3. Berners-Lee, Tim, James Hendler, and Ora Lassila. "The Semantic Web," *Scientific American*. New York: Scientific American, May 2001.

uses a unique configuration. If equipment previously installed at the customer's site has reported a particular maintenance condition to your Semantic Web, such as "upgrade to new OS," the site will return download information for the latest version.

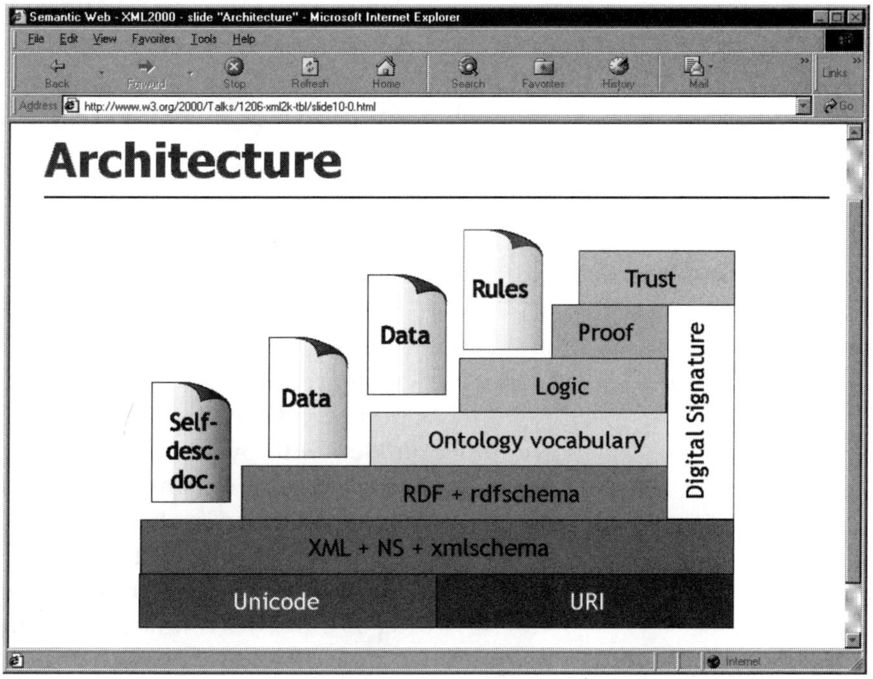

Figure 10-1. The Semantic Web builds meaning into the Internet using XML and RDF. (World Wide Web Consortium, www.w3.org/2000/Talks/1206-xml2k-tbl/slide10-0.html)

According to Berners-Lee, the biggest opportunity for the Semantic Web will be realized when software agents begin to exchange results with each other. Their effectiveness will increase dramatically as more machine-readable Web content becomes available. Think of the possibilities for software upgrades alone. In the past, when a version 4 HMI software package came across a version 5 document, it said it couldn't read it. The program gave up because it assumed it couldn't possibly understand something that didn't exist when it was developed. But with RDF and the inference languages that are associated with it, a version 5 document will be self-describing. It will provide a Web-based link to its own information, which the version 4 document will use to read the steps it needs to follow to convert a version 5 document to version 4. The only requirement is that the version 4 document must have been written so that it can understand the language in which the rules were written. That is the function of RDF.

In the realm of logistics, imagine that the software agent that handles your calendar has the capacity to coordinate with the agent that handles your travel. The goal is to arrange a series of technical briefings about your newest product upgrade with the engineering team of your top ten customers. If all this were handled automatically, online, with minimal human intervention, think of the time that you and your customers would save and the frustration you would avoid.

Of course, we've been hearing about the imminent arrival of artificial intelligence and software agents for fifteen years. With the exception of a few primitive programs that can help customers configure mildly complex products, we have yet to see the technology move beyond the research stage. But it will.

Software Agents

In fact, the Internet is already crawling with Web robots ("bots"), which are the first real precursors of intelligent agents. As we saw in Chapter 5, bots are software designed to help people deal with overwhelming amounts of information. They arrived on the scene a few years ago at the same time as the first Internet search engines. Commonly referred to as spiders or Web crawlers, they explore the Web by requesting documents from Web servers, examining them for keywords or embedded links, and then sending out requests for new URLs so they can retrieve new files or documents.

Search bots, for instance, are software programs used by commercial search engines to monitor Web sites and rank them according to such criteria as keyword "hits" and the number of links from other sites. The most well known of the search bots is "Scooter," the brains behind the AltaVista search engine. AltaVista is one of largest search engines on the Net with over three hundred million pages in its huge database. Most search engine bots are having a tough time keeping up with the Internet these days; some take as long as three months to reindex changes to a site. But Scooter has carried its weight, collecting data from more than three million Web pages a day. Its sister software, NetIndexer 2, indexes more than one gigabyte of data per hour.

Shopping bots are also crawling the Net to help buyers look for the best deals on a variety of products, mostly consumer goods like CDs and toys. However, many companies refuse to compete strictly on price and thus block shopping bots from searching the content of their Web sites. One of the second-generation shopping bots, developed by Frictionless (www.frictionless.com), has since been converted into a Java/XML-based suite of Web-based procurement services. The use of bots for business and supply chain management seems to have won out over consumer bots.

Other bots include "chatter bots," that is, software programs that answer frequently asked questions, and financial bots. In the latter category, CompanySleuth.com's bot scours the Net for business and financial news about public companies and reports this information to its subscribers, most of whom are independent investors.

Data Mining

With the exception of search engine bots, most bots do not troll the entire Internet. Instead, they scour only the major search engines or select USENET newsgroups. Industry insiders believe that the next generation of bots will be smarter and more comprehensive. When searching news stories, for instance, some bots grab entire stories but most retrieve only a summary or provide a link to the full story.

Bots have great potential in the area of data mining, which is the process of finding patterns in large amounts of data. Because they refine their efforts as they go, they can sort through gigabytes of data, saving many hours of human labor. Many shopping sites use artificial intelligence techniques to learn the interests of shoppers and thus provide a more personalized experience. Amazon.com, for instance, keeps track of the items you research and purchase. Over time, it updates your profile and delivers relevant content in the form of recommendations for future purchases. At the moment, such personalization technology is quite expensive. More rudimentary capabilities are offered by personalization software, which relies on the visitor to the site to fill in forms that specify his or her preferences.

Biometrics

Security remains a central issue for the Internet. Individual users are concerned about such privacy issues as the amount and type of information being gathered about them and their online habits. With the rapid growth of e-commerce, authentication during online shopping has become a major issue. Is the credit card owner actually making the purchase? Is a legitimate employee of the vendor or credit card company accessing the customer's information? With the growth of the telecommuting workforce, how does one allow selective and secure access to sensitive corporate or government data?

As the information age continues to sweep through our lives, there is no doubt that new needs involving privacy issues both online and in the physical world have arisen. In the past decade, a new science called biometrics has come to dominate developments in the security field. As a result, biometrics has been transformed from the general application of statistics to biological observations to a more narrowly focused science that addresses security concerns through personal identification. A new repertoire of identification and verification techniques, such as voice printing, retinal scans, photo imaging, digital signatures, finger imaging, smart cards, face recognition, even DNA matching, is emerging as quickly as new CPUs can keep up with the number-crunching demands of the algorithms.

One of the most popular biometric applications is fingerprint recognition. This involves scanning a fingertip optically or digitally, entering a digitized record in a computer database, and comparing the finger image to an original for verification or identification purposes. Identifying individual finger patterns entails an analysis of small ridge endings on the finger, known as minutiae. The padded part of the average finger contains between forty and sixty minutiae, which can be analyzed as to location and position. Because each fingerprint is unique, the pattern of a person's minutiae is also unique. To speed the process, designers of fingerprint recognition systems tend to limit the number of minutiae points analyzed to a few small areas within the print.

Every day, biometrics advances further and infiltrates our lives just a bit more. It promises to make our computer networks more secure and reduce the theft of information and corporate assets. For example, SecuGen Canada (www.secugen.com) recently announced that the Credit Union Central (CUC) of British Columbia is using fingerprint recognition to authenticate all access to its central security portal. The portal is a gateway to a large array of financial applications used by CUC's membership of Canadian credit unions and corporate clients.

Biometric devices are becoming more ergonomic and practical, as exemplified by the computer mouse that recognizes fingerprints. Many leading developers are working to integrate biometric functionality into our everyday computing lives. Intel and Compaq, for instance, are active members of the BioAPI Consortium (www.bioapi.org), and Microsoft is working to integrate biometrics into future versions of Windows. Even notebook and handheld computers will soon incorporate biometrics into integrated multimedia devices, including Webcams and microphones.

The Future of Manufacturing

Regardless of the Internet's security issues, the manufacturing process will continue on, struggling to meet ever-increasing demands for higher volume, better quality, and lower costs. For decades, the dream of manufacturers has been "build to order" (BTO)—that is, carry no inventory and manufacture only what customers want and are willing to pay a fair price for. But assembling complex products is challenging enough without requiring a tooling change after each unit is made. As a result, successful BTO implementations are few and far between.

Thus arose "mass customization," that is, the manufacture of highly individuated products on a mass scale. In the past decade, customization of this kind was handled by highly paid salespeople working closely with their customers. Companies like Motorola's paging division and Dell Computer were front-runners in developing a vast network of resellers and call centers to run what are essentially order fulfillment operations. Premanufactured components are assembled and shipped according to specific configurations.

Several years ago, Dell took the next step by replacing its reseller network with an interactive Web site. The purpose of the site was twofold: to empower customers to configure a custom computer system and to shorten the supply chain. And it's worked. With an average delivery time to customers of six days, Dell has shaken the computer industry to its foundation. Now it is fine-tuning the process to speed up its order, fulfillment, shipping, and payment operations. As shown in Figure 10-2, automotive factories must provide a flexible supply chain, as well, if they hope to make any inroads into the "build-to-order" methodology.

Globally Distributed Processes

In *The Technology Machine*, Patricia Moody and Dick Morley describe the Web-based manufacturing efforts of contract manufacturer EFTC (www.eftc.com). That company is using the Internet to tighten its supply chain by electronically partnering with transportation providers and component distributors. EFTC can now boast of a build-to-order infrastructure that minimizes time and costs by locating its key services in central hubs close to key customer locations. Moody and Morley state that "EFTC's pioneering collaboration with Fujitsu and logistics providers to repair, return, and even assemble electronics devices, without the Colorado plant ever touching the product, makes the point that

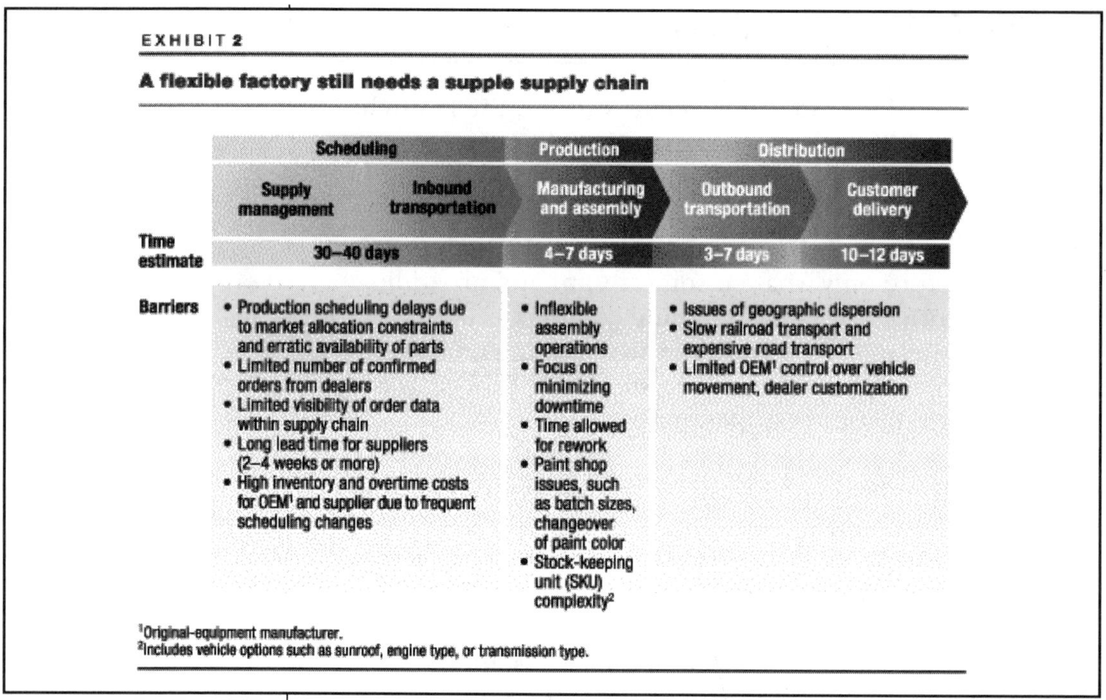

Figure 10-2. A flexible automotive factory still needs a nimble supply chain to make any kind of significant inroads into build to order processes. (Mani Agrawal, T.V. Kumaresh, and Glenn Mercer, "The False Promise of Mass Customization," The McKinsey Quarterly, 2001, Number 3, www.mckinseyquarterly.com. Used by permission.)

innovative logistics solutions are not all about where to put the warehouse or which air freight company gets there faster. It's about who does the work and who captures the money."

Michael Dertouzos, former director of the MIT Laboratory of Computer Science, agrees that the trend toward minimal inventory will mean that most manufacturing will take place in a variety of locations, close to the product's final destination. As an example, he outlines the process for buying your dream car in the near future.[4] Data-mining software running at the corporate headquarters determines broad purchasing trends based on inputs from customer-focused Web sites and dealers. Based on this information, premanufactured components are ordered when needed. The pieces are delivered to the local dealer who assembles the car that you custom ordered just the week before.

Sounds wonderful. Unfortunately, as Dertouzos points out, the automotive manufacturing industry isn't set up to handle such a widely distributed organization. What would become of the hundreds of robots currently lining the plant floor? Without these robots, who will perform the thousands of welding and painting operations? The answer might be

4. Michael Dertouzos, *What Will Be.* San Francisco: HarperEdge, 1997, p. 196.

that a new type of composite material will be developed that enables a car to be snapped together using a dozen or so prefabricated assemblies. If it can work for a house, why not a car?

However auto assembly evolves in the future, manufacturing is at a crossroads today. High-speed distributed computer technology, software agents, open-source components, and the Internet are pushing manufacturing toward a new future (see Sidebar "Automation Industry Begins to Turn to Open-source Systems as Path to Future"). In the words of Moody and Morley, "Our society has become consumption constrained as opposed to production constrained. The result is that the way we think about production and consumption will change. We will locate innovative replication facilities at the point of consumption, in quantities of one or one hundred, with zero defects, with zero inventory in process. Desktop manufacturing, virtual prototyping, and the Internet are the primary enabling technologies."[5]

Privacy and the Human Element

Overriding all the new technologies, from the Semantic Web to biometrics, is the concern about the invasive nature of the Internet. Already, "identity theft" crimes have increased since the advent of the Web. Government agencies and commercial organizations must work together to ensure a satisfactory quality of life for employees and customers in our new e-business future.

5. Patricia E. Moody and Richard E. Morley. *The Technology Machine.* New York: Free Press, 1999, p. 295.

Automation Industry Begins to Turn to Open-source Systems as Path to Future

by Ken Crater, President, Control.com Inc.

When Control.com Inc. launched its new business in 1999 (see Figure 10-3), we took a cue from our mission to provide open industrial control systems and decided to base many of our internal systems on open software products. The result has, on balance, been a significant advantage.

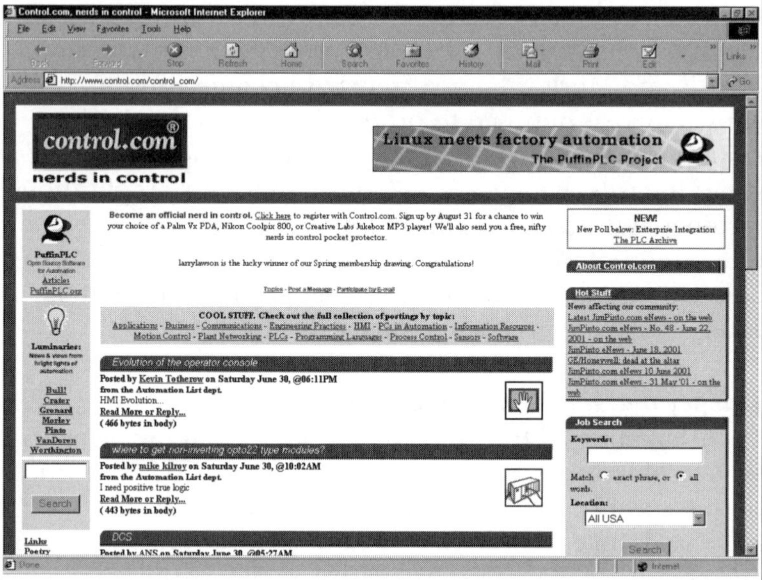

Figure 10-3. Control.com relies on open source components to run its innovative community for control engineers.

Part of our business strategy is the development of a large online community of automation professionals. This required a sophisticated infrastructure of computers, database-driven Web tools, and server software, all of which we chose from a palette of open-source solutions.

For our server operating system, one of the key criteria was a platform that supported the range of software we require for our business. In today's open-source world, the least common denominator is Linux. But which of the various releases of Linux should we choose? Although each Linux variant has its (sometimes passionate) adherents, we determined that Red Hat Linux (www.redhat.com) is enough of a market leader that many of the software products we needed had already been tested and packaged for it. Red Hat Linux became our server platform of choice.

Although Red Hat Linux, as with all versions of Linux, is freely downloadable on the Internet, we paid a nominal amount for our copy to obtain a limited amount of support from Red Hat. Note, however, that as we started to use the operating system on additional machines for testing and experimentation, we were not obligated by the terms of the license to purchase additional copies since the purchase price is for support, not the software itself.

Flexibility

To implement our Web-based community, we needed a flexible system for entering, storing, sorting, and displaying messages created by the community members. We were intrigued by the early work of a company called Digital Creations on their Web development platform called Zope (www.zope.com). This product stored messages along with related data in an object-oriented database. Further, all of the Web pages designed for accessing that data are also stored in the database, which provided features such as the ability to incrementally undo design changes made to the site.

(continued)

Better still, Digital Creations had recently made the decision to release Zope as an open-source product. There were two implications of this decision for our company. The first is that we could download and work with the product for free. The second, and ultimately much more important, reason is that we had complete access to the source code for the product. Thus, we could troubleshoot and resolve any problems we encountered or even create a customized version for our own application.

The Value of Integrated e-Mail

Another aspect of our use of open software is still evolving, and with an interesting twist. Historically, we had been using a commercial e-mail list server product that ran under Windows NT for our previous electronic community. As we developed additional e-mail lists for subcommunities, we began to deploy an open-source list manager called *Mailman*. Our ultimate goal is the convergence of our e-mail-based communities with our Web-based communities, to make life richer for our users and easier for our administrators. By a stroke of luck, Digital Creations hired the developer who originally wrote Mailman, setting the stage for a commercially supported integration of the product into Zope.

The question, however, was "when?" Open-source projects tend to have a mind of their own. Rarely subject to a schedule, they progress according to the whims or needs of the authors, and new versions are released "when they are ready." Although Digital Creations is still interested in the evolution of Zope, their interests are tied to their own business use of the product, rather than to the needs of companies using the product for free. Is this a dilemma?

As it turns out, the answer is "no." Since one of the primary interests of any business is making money, the key to getting our business needs met was evident: pay Digital Creations to make our changes. For a fraction of what such a project would normally cost, they agreed to accelerate their plans to integrate Zope and Mailman while taking into consideration the specific needs of our Web strategy. They were happy to do this since it amounted to a subsidy of their future plans anyway. We were happy to do this since we were getting a customized product that is destined to become the next version of a standard—and free—product.

But even if this happy coincidence had not occurred, we would still have had the option of modifying the code ourselves or hiring an independent consultant to do so. For any business whose strategies have been held hostage by a software vendor's development priorities, that is a compelling advantage indeed.

In the End...

We are on the verge of seeing some incredible changes in the way we do business. The industrial enterprise of the future will combine new business practices with new technology to deliver information and products at incredible speeds:

- Wireless technology will combine with the Internet to deliver "any-to-any computing." You will soon be able to get to any corporate data from anywhere by interacting with the device of your choice (see Figure 10-4).
- Embedded computing will merge with the Internet and lead to truly intelligent factory-floor devices: self-

inspecting, self-diagnosing, but above all, globally communicating.

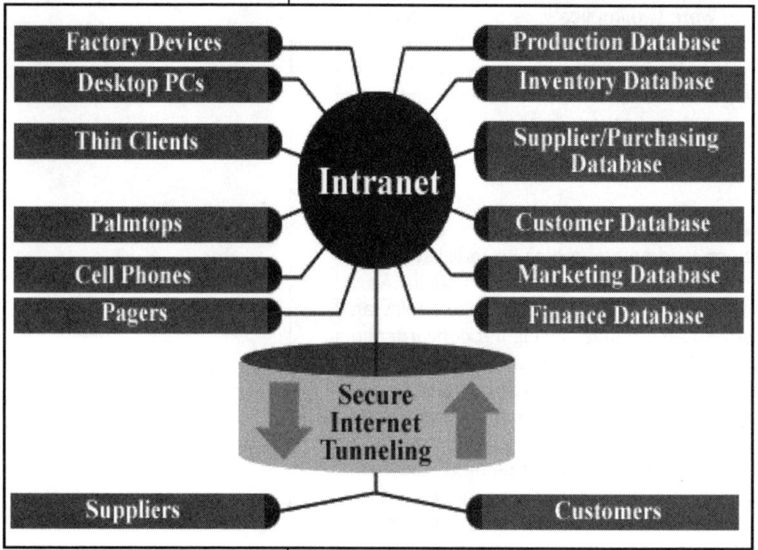

Figure 10-4. In the factory of tomorrow, wireless technology will interface anyone anywhere to intelligent factory devices; distributed databases will empower customers and build loyalty; and the high speed Internet will connect us to the world.

- Software agents will fuse with the Internet to deliver a new level of automatic, computer-generated interactions. These functions will be carried out with a minimum of human involvement, thus freeing customers and employees from many of our more mundane tasks. Such convenience will serve to bind customers more closely to those vendors who provide such indispensable tools.

The Internet is a revolutionary tool for empowerment. Customers, factory-floor workers, field technicians, even plant-floor devices are already being affected by it. e-Business is here to stay. Make sure your enterprise is making the most of it.

Selected Readings, Glossaries, & Online Resources

It's nearly impossible to keep up with all that's happening on the Internet these days. We recommend you start with the following books and articles produced by an impressive group of industry experts. The Internet itself also contains a wealth of resources so we've assembled a list of relevant Web sites where you can start your search.

Readings

Berners-Lee, Tim, James Hendler, and Ora Lassila. "The Semantic Web," *Scientific American.* New York: Scientific American, May 2001.

Berners-Lee, Tim. *Weaving the Web.* San Francisco, California: HarperSanFrancisco, 1999.

Dertouzos, Michael. *The Unfinished Revolution: Human-Centered Computers and What They Can Do for Us.* New York: HarperCollins Publishers, 2001.

Dertouzos, Michael. *What Will Be: How the New World of Information Will Change Our Lives.* San Francisco, California: HarperEdge, 1997.

Dick, Kevin. *XML: A Manager's Guide.* Reading, Massachusetts: Addison-Wesley, 2000.

Fingar, Peter, Harsha Kumar, and Tarun Sharma. *Enterprise e-Commerce.* Tampa, Florida: Meghan-Kiffer Press, 2000.

Gilder, George. *Telecosm: How Infinite Bandwidth Will Revolutionize Our World.* New York: The Free Press, 2000.

Iacobucci, Dawn, Editor. *Kellog on Marketing: The Kellogg Marketing Faculty, Northwestern University.* Chicago IL, John Wiley and Sons Inc., 2000.

Kalakota, Dr. Ravi and Marcia Robinson. *e-Business 2.0: Roadmap for Success.* Reading, Massachusetts: Addison-Wesley, 2001.

Kelly, Kevin. *Out of Control: The New Biology of Machines, Social Systems, and the Economic World.* Reading, Massachusetts: Addison-Wesley, 1994.

Leonard, Andrew. *Bots: The Origin of New Species.* San Francisco, California: Hardwired, 1997.

Levine, Rick, Christopher Locke, Doc Searls, and David Weinberger. *The Cluetrain Manifesto.* Cambridge, Massachusetts: Perseus Publishing, 2000.

Linthicum, David S. *B2B Application Integration: e-Business-Enable Your Enterprise.* Reading, Massachusetts: Addison-Wesley, 2001.

Micklethwait, John and Adrian Wooldridge. *A Future Perfect: The Challenge and Hidden Promise of Globalization.* New York: Crown Business, 2000.

Moody, Patricia E. and Richard E. Morley. *The Technology Machine: How Manufacturing Will Work in the Year 2020.* New York: The Free Press, 1999.

Peppers, Don and Martha Rogers, Ph.D. *The One to One Manager: Real World Lessons in Customer Relationship Management.* New York: Currency-Doubleday 1999.

Riddersråle, Jonas and Kjell Nordström,. *Funky Business: Talent Makes Capital Dance.* Stockholm, Sweden: ft.com, 2000.

Seybold, Patricia B. with Ronni T. Marshak. *Customers.com: How to create a profitable business strategy for the Internet and beyond.* New York: Times Business Press, 1998.

Siebel, Thomas M. *Taking Care of eBusiness.* New York: Currency Doubleday, 2001.

Treese, G. Winfield and Lawrence C. Stewart. *Designing Systems for Internet Commerce.* Reading, Massachusetts: Addison-Wesley, 1998.

Trout, Jack with Steve Rivkin. *Differentiate or Die: Survival in Our Era of Killer Competition.* New York, NY: John Wiley and Sons Inc., 2000.

Werth, Jacques and Nicholas E. Ruben. *High Probability Selling.* Dresher, PA: Abba Publishing, 1999.

Wurman, Richard Saul. *Information Anxiety 2.* Indianapolis, Indiana: Que, 2001.

Online Glossaries

ASPstreet – www.aspstreet.com

Business 2.0 – www.ecompanynow.com/glossary

Darwin Online – www.darwinmag.com/index.html

Electric Commerce Dictionary – www.tedhaynes.com/haynes1/atol.html

Gartner Group – gartner11.gartnerweb.com/btj/static/glossary/glossary.html

Logistics – theweb.badm.sc.edu/sealandx/logisticsg1.htm

Quality – theweb.badm.sc.edu/sealandx/qualityg.htm

Net Market Makers – www.netmarketmakers.com/glossary

Manufacturing Systems – www.manufacturingsystems.com/glossary

Merriam-Webster Online – www.m-w.com/dictionary.htm

Public Procurement and Purchasing – www.euro profile.co.uk/glossary.htm

Purchasing Terms – www-storehouse.ucdavis.edu/preference/glossary.html

TechWeb Encyclopedia – www.techweb.com/encyclopedia

Webopedia – www.webopedia.com

Whatis.com – whatis.techtarget.com

Workz.com glossary – www.workz.com/content/1505.asp

MES and Supply Chain Integration Links

MESA International – www.mesa.org

UM/Rolla CIM Lab – www.umr.edu/~emgt/ciml/index.html

Computerworld MES primer – www.computerworld.com/cwi/story/0,1199,NAV47-74-212-466_STO47639,00.html

McGill University CIM Lab – www.cim.mcgill.ca/~ialab/english/english.html

e-Optimization – www.e-optimization.com

Institute for Operations Research and Management Sciences – www.informs.org

The CRM Project – www.crmproject.com/documents.asp?d_ID=773

Sybase – www.sybase.com/detail/1,3693,1012546,00.html

University of Maryland Supply Chain Management Center – www.rhsmith.umd.edu/scmc

Stanford Supply Chain Forum – www.stanford.edu/group/scforum/

University of Tennessee Knoxville – thecenter.utk.edu/programs/integrated-supply-chain-certification.html#integrative_supply_chain_experience

Michigan State University Supply Chain Management Center – www.bus.msu.edu/msc/scm

APICS- The Educational Society for Resource Management – www.apics.org

National Association of Purchasing Management – www.napm.org

Other Links

Advanced Integrated Manufacturing Center (AIM) – www.aimcenter.org

Automotive Industry Action Group Online – www.aiag.org/links

American Society of Mechanical Engineers (ASME) – www.asme.org

American Society of Professional Estimators (ASPE) – www.aspenational.org

American Society for Quality (ASQ) – www.asq.org

American National Standards Institute – www.ansi.org

Avraham Y. Goldratt Institute – www.goldratt.com

American Textile Manufacturers Institute (ATMI) – www.atmi.org

Best Manufacturing Practices – www.bmpcoe.org/index.html

BusinessWeek – www.businessweek.com

Brainstorm Group, Inc. – www.brainstorm-group.com

CARGOLOG Transportation and Logistics Directory – www.cargolog.com

CEOexpress – www.ceoexpress.com/default.asp

Decision Support Systems Resources – dssresources.com

Distribution Business Management – www.dcenter.com

Georgia Institute of Technology – www.emil.gatech.edu/index.cfm

Georgia Tech Economic Development Institute – cmit.edi.gatech.edu

ISA—The Instrumentation, Systems, and Automation Society – www.isa.org

International Society of Logistics (SOLE) – www.sole.org

LORS Library of Online Resources on Scheduling – ie.bilkent.edu.tr/~lors

Michael Trick's Operations Research Page – mat.gsia.cmu.edu

National Association of Manufacturers – www.nam.org

National Institute of Standards and Technology Manufacturing (NIST MEP) – www.mep.nist.gov

Shingo Prize for Excellence in Manufacturing – www.shingoprize.org/shingo/index.html

Society of Manufacturing Engineers (SME) – www.sme.org

ThomasRegional.com – www.thomasregional.com

Tooling and Manufacturing Association (TMA) – www.tmanet.com

UK Planning and Scheduling Special Interest Group – www.research.salford.ac.uk/plansig/home.html

University of Pennsylvania/Wharton School of Business – www.wharton.upenn.edu

Wisconsin Manufacturing Extension Partnership – www.wmep.org

York Technical College – www.yorktech.com/Home/IMP.htm

Security References

Web Sites

CERT Coordination Center (CERT/CC) – www.cert.org

Computer Security Resource Center – csrc.nist.gov

Cybercrime.gov – www.cybercrime.gov

High Technology Crime Investigation Association (HTCIA) - www.htcia.org

International Information Systems Security Certification Consortium (ISC2) – www.isc2.org

National White Collar Crime Center – www.nw3c.org

American Society for Industrial Security (ASIS) – www.asisonline.org

Applied Computer Security Associates (ACSE) – www.acsac.org

Center for Education and Research in Information Assurance and Security (CERIAS) (Purdue Univ.) – www.cerias.purdue.edu

Center for Secure Information Systems (CSIS) – www.isse.gmu.edu/~csis

Computer Incident Advisory Capability (CIAC) – www.ciac.org/ciac

Computer System Security and Privacy Advisory Board (CSSPAB) – csrc.nist.gov/csspab

Critical Infrastructure Assurance Office (CIAO) – www.ciao.gov

Government Information Technology Services (GITS) – www.ecommerce.gov

Information on Computer Security (U. Berkeley) – http.cs.berkeley.edu/~daw/compsec.html

Information on Cryptography (U. Berkeley) – http.cs.berkeley.edu/~daw/crypto.html

Information Systems Audit and Control Association (ISACA) – www.isaca.org

Information Systems Security Association Inc. (ISSA) – www.issa-intl.org

International Association for Computer Systems Security Inc. (IACSS) – www.iacss.org

International Society for Professionals in E-Commerce (iSPEC) – www.ispec.org

Internet Fraud Complaint Center (IFCC) – www.ifccfbi.gov

Kerberos: The Network Authentication Protocol (MIT) – web.mit.edu/kerberos/www

The IT Governance Institute – www.itgovernance.org

National Classification Management Society Inc. – www.classmgmt.com

National Coalition for the Prevention of Economic Crime – www.ncpec.org

National Information Assurance Partnership (NIAP) – www.niap.nist.gov

National Security Institute – www.nsi.org

National Security Telecommunications and Information Systems Security Committee (NSTISSC) – www.nstissc.gov

Network Security Framework Forum (NSFF) – www.nsff.org

PKI Forum – www.pkiforum.org

Ronald L. Rivest: Crytpography and Security (MIT) – theory.lcs.mit.edu/~rivest/crypto-security.html

University of California at Davis – seclab.cs.ucdavis.edu

Online Magazines and Newsletters

2600 – www.2600.com

Crypto-Gram Newsletter – www.counterpane.com/crypto-gram.html

Cryptome – www.cryptome.org

Electronic Privacy Information Center (EPIC) – www.epic.org

ICSA – www.trusecure.com

Information Security Magazine – www.infosecuritymag.com

Information Security News – www.infosecnews.com

Internet Security Review Online – www.isr.net

Mountain Wave Inc. - Computer Security News Daily – www.mountainwave.com

Security Focus – www.securityfocus.com

Security Magazine – www.securitymagazine.com

Security Portal – www.securityportal.com

Security Watch – www.securitywatch.com

Index